ISBN 978-1-332-15280-3
PIBN 10291750

1 MONTH OF
FREE
READING

at

www.ForgottenBooks.com

By purchasing this book you are eligible for one month membership to ForgottenBooks.com, giving you unlimited access to our entire collection of over 1,000,000 titles via our web site and mobile apps.

To claim your free month visit:
www.forgottenbooks.com/free291750

English
Français
Deutsche
Italiano
Español
Português

www.forgottenbooks.com

Mythology Photography **Fiction**
Fishing Christianity **Art** Cooking
Essays Buddhism Freemasonry
Medicine **Biology** Music **Ancient
Egypt** Evolution Carpentry Physics
Dance Geology **Mathematics** Fitness
Shakespeare **Folklore** Yoga Marketing
Confidence Immortality Biographies
Poetry **Psychology** Witchcraft
Electronics Chemistry History **Law**
Accounting **Philosophy** Anthropology
Alchemy Drama Quantum Mechanics
Atheism Sexual Health **Ancient History**
Entrepreneurship Languages Sport
Paleontology Needlework Islam
Metaphysics Investment Archaeology
Parenting Statistics Criminology
Motivational

THE LOST TOWNS OF THE
YORKSHIRE COAST

PRINTED AT BROWNS' SAVILE PRESS,
SAVILE STREET AND GEORGE STREET, HULL.

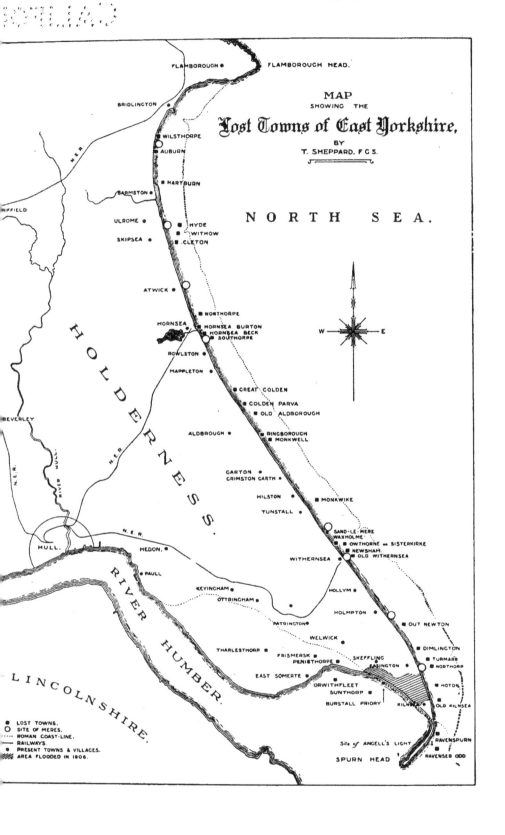

MAP
SHOWING THE

Lost Towns of East Yorkshire,
BY
T. SHEPPARD, F.G.S.

NORTH SEA.

FLAMBOROUGH ●   FLAMBOROUGH HEAD.

BRIDLINGTON ●

WILSTHORPE ●
○ AUBURN

■ HARTBURN

BARMSTON ●

ULROME ●   ○
○ ■ HYDE
SKIPSEA ●   ■ WITHOW
■ CLETON

ATWICK ●   ○

■ NORTHORPE
HORNSEA   ■ HORNSEA BURTON
○   ■ HORNSEA BECK
■ SOUTHORPE
ROWLSTON ●

MAPPLETON ●

■ GREAT COLDEN

■ COLDEN PARVA
■ OLD ALDBOROUGH

ALDBROUGH ●   ■ RINGBOROUGH
■ MONKWELL

BEVERLEY ●

GARTON ●
GRIMSTON GARTH ●

HILSTON ●   ■ MONKWIKE
TUNSTALL ●

○ SAND-LE-MERE
WAXHOLME ■
■ OWTHORNE or SISTERKIRKE
■ NEWSHAM.
WITHERNSEA ●   ■ OLD WITHERNSEA

HULL ○

HEDON ●

● PAULL

KEYINGHAM ●
OTTRINGHAM ●

HOLLYM ●

HOLMPTON ●

PATRINGTON ●   ■ OUT NEWTON

WELWICK ●
THARLESTHORP ●   ■ DIMLINGTON
FRISMERSK ●   ■ TURMARR
PENISTHORPE ■   SKEFFLING ●   ■ NORTHORP
EASINGTON ●   ○
EAST SOMERTE ■   ■ HOTON
ORWITHFLEET ●
SUNTHORP ●   KILNSEA ●   ■ OLD KILNSEA
BURSTALL PRIORY ●

■ RAVENSPURN
Site of ANGELL'S LIGHT   ■
SPURN HEAD.   RAVENSER ○○○

LINCOLNSHIRE.

RIVER HUMBER.

HOLDERNESS.

RIVER HULL

■ LOST TOWNS.
○ SITE OF MERES.
‥‥ ROMAN COAST-LINE.
━ RAILWAYS.
● PRESENT TOWNS & VILLAGES.
▨ AREA FLOODED IN 1906.

THE LOST TOWNS OF THE YORKSHIRE COAST

AND OTHER CHAPTERS BEARING UPON THE GEOGRAPHY OF THE DISTRICT

BY

THOMAS SHEPPARD

F.G.S., F.R.G.S., F.S.A.(Scot).

AUTHOR OF "GEOLOGICAL RAMBLES IN EAST YORKSHIRE,"
"THE EVOLUTION OF HULL," "MAKING OF EAST
YORKSHIRE," ETC., ETC.
EDITOR OF MORTIMER'S "FORTY YEARS' RESEARCHES,"

LONDON

A. BROWN & SONS, LIMITED, 5 FARRINGDON AVENUE, E.C.
HULL: SAVILE STREET   YORK: CLIFFORD STREET

—

1912

DA6
Y6S5

TO

THE RIGHT WORSHIPFUL

THE MAYOR OF KINGSTON-UPON-HULL

(ALDERMAN JOHN BROWN),

WHO HAS DONE SO MUCH TO PRESERVE THE VALUABLE

HISTORICAL EVIDENCES IN HULL

AND THE SURROUNDING

DISTRICT

261616   *a 2*

PREFACE

From his preface to " The Lost Towns of the Humber," it is apparent that the intention of the late J. R. Boyle was to publish a companion volume on the " Lost Towns and Churches of the Yorkshire Coast "; in fact, I am given to understand the first chapter was actually put into type, and the frontispiece (Hornsea Church) was engraved for the work, but it got no further. The proof of the chapter was lost, and I regret I never saw it.

Upon many occasions I urged Mr. Boyle to complete the work, but as circumstances have prevented this, I have ventured to do it myself. I am well aware of the fact that the present work must lack much that would characterise it were it prepared by the author of " The Lost Towns of the Humber," but there is a considerable amount of information now available which may not be so readily accessible to future workers. I am familiar with many aspects of our coast that, through the ravages of the sea, will be denied to succeeding workers. A number of cliff measurements, documents, plans and charts have also, in various ways, come under my notice, many of which have not been previously known.

Further, in the hope that the book may be of service to students, chapters have been added referring to the

PREFACE

geology, antiquities, natural history, and other subjects
which, now-a-days, are included under the comprehensive
head of " geography."

Most of the illustrations have been made specially
for this work. A few, however, have been taken from
previous papers of my own, and I am indebted for the
loan of blocks to the Yorkshire Archæological Society,
the Yorkshire Philosophical Society, the East Riding
Antiquarian Society, the Hull Geological Society, the
Yorkshire Geological Society, and the Hull Scientific
Club. To many friends, also, who are duly named,
I am indebted for the loan of photographs from which
some of the illustrations have been made. I have
also received help in the matter of proof-reading, etc.,
from Mr. T. Stainforth, B.A.

Municipal Museums,
Hull, *Nov.* 1912.

CONTENTS

CHAPTER I

CHAPTER II

CHAPTER III

CHAPTER IV

CHAPTER V

CHAPTER VI

CHAPTER VII

CONTENTS

CHAPTER VIII

CONTENTS

CHAPTER XVII

CHAPTER XVIII

CHAPTER XIX

CHAPTER XX

CHAPTER XXI

CHAPTER XXII

CHAPTER XXIII

CHAPTER XXIV

CONTENTS

CHAPTER XXV

CHAPTER XXVI

CHAPTER XXVII

CHAPTER XXVIII

CHAPTER XXIX

CHAPTER XXX

CHAPTER XXXI

CHAPTER XXXII

CHAPTER XXXIII

LIST OF ILLUSTRATIONS

LIST OF ILLUSTRATIONS

LIST OF ILLUSTRATIONS

LIST OF ILLUSTRATIONS

THE LOST TOWNS OF THE YORKSHIRE COAST

CHAPTER I

INTRODUCTION—ROYAL COMMISSION ON COAST EROSION—
MAGAZINE LITERATURE—BRITISH ASSOCIATION RE-
PORTS—RATES OF COAST EROSION—DEEP SEA
EROSION.

THERE can be no question that the changes on the
south-east coast of Yorkshire have been as great as those
in any part of Britain, and in various ways geologists
and antiquaries have placed on record a mass of information relative to these changes. The result is that, probably
more than is the case with any other district, we are able
to get reliable information in reference to one of the most
interesting chapters in the geography of our islands.
In the following pages an attempt has been made to
summarise the information to be found in the scores upon
scores of papers and documents, and in addition a few
chapters have been added bearing upon the more modern
aspects of the geography of the East Riding, a district
which, in the writer's opinion, is second to none for
geological and antiquarian interest.

That the sea has been encroaching upon the land is
well-known, and the voluminous reports of the recent

Royal Commission on Coast Erosion (one of which alone contains over a thousand closely-printed folio pages), will speak fairly forcibly as to the importance of the subject.

Fortunately, in East Yorkshire, with the aid of the work and publications of the various local geological, natural history, and antiquarian societies, the large number of valuable memoirs by geologists and antiquaries of the first rank (some of which date back a century and more ago), and by the aid of an unusually complete series of maps, charts, and engravings, it has been possible to give a narrative of exceptional completeness. Of all these, full advantage has been taken by the present writer, and in addition, he has been able to include a number of his own observations, made during very frequent peregrinations along this most interesting coastline.

According to the Reports of the Royal Commission already referred to, it seems that between 1848 and 1893, 774 acres were lost in Yorkshire, principally from Holderness. On the other hand, during the same period, 2178 acres have been reclaimed within the Humber estuary, so that really our county has considerably increased in size during that period.

Unquestionably whole towns and villages have been washed away, and the loss of land and property is still going on.

It goes without saying, however, that the reports are frequently over-estimated. Perhaps the gem of these appeared in a " popular " magazine, dated April 7th, 1906. In this it was stated :—" On the coast of Yorkshire there are two noses of hard rock that the sea can eat but slowly. They are Flamborough Head and Spurn Point [!], and between them lie 33 miles of coast, which the North Sea is swallowing at the rate of 3 yards in every 12 months. At Withernsea, just to the north of Spurn Point, houses *go over the cliff almost daily* [!]. Some little time ago, there lived at Withernsea, an old fisherman who, despite the warnings of his friends, persisted in declaring that the sea would never harm him or his. . . . There were two houses between the old fellow's cottage and the crumbling cliff-edge. . . . One rough night, however, a biting nor'-easter hurled the ramping breakers against the shore to such purpose that first one house went, and then the other. Then the wall of the old fisherman's cottage collapsed, because of the disturbance to the foundations ; and he awoke in the grey of the morning to find himself looking straight from his bed on to the green waters of the North Sea."

All this is a delightful piece of " Home Chat," which is probably the product of the same inventive brain that tells us there are only *two* noses of *hard rock* in Yorkshire, and one of them is Spurn Point!

As regards the quantity of material removed, Mr. E. R. Matthews writes : *—" Assuming the average height of the cliff to be 18 feet, and the weight per cubic foot of the glacial deposit and boulder clay to be 120 lbs., no

* Proc. Inst. Civil Eng. 1905.

less than 1,904,194 tons of cliff are washed away annually.
. . . Assuming that the rate of erosion of 3 yards per
annum has been going on for the past century, no less
than 3,052 millions tons of cliff have been swallowed up
during that time. As to area lost, assuming that one
acre per mile of coastline has been swallowed up annually,
and that loss at this rate has been continuous, no less
than 66,600 acres have been swallowed up by the sea
since the time of the first Roman Invasion. . . . If
the coastline between Flamborough Headland and Bridlington—5 miles in length—be added, and the rate of
erosion of that part of the coast be taken as 6 feet per
annum, the loss is found to be 32 acres per annum, or
7,180 acres swallowed up in the period named. This
makes a total of 73,780 acres, or 115 square miles of land
since the Roman Invasion, between Flamborough Head
and Kilnsea, . . . an area almost equal to that
upon which London stands. . . . The sea has made
an inroad upon the land since Roman times of $3\frac{1}{2}$ miles."

The British Association for the Advancement of
Science in 1884 appointed a Committee to " enquire into
the Rate of Erosion of the Sea-coasts of England and
Wales, and the influence of the Artificial Abstraction of
Shingle or other material in that action." The Committee
sat for twelve years, and published four reports. In
the report presented at Ipswich in 1895 is a memorandum from the Director-General of the Ordnance Survey,
based upon the two surveys of 1852 and 1889 respectively.
Obviously these reports are as reliable as is possible

for reports to be, so I take the liberty of quoting from them fairly fully :—

Capt. A. H. Kennedy (1890) reports :—" The following measurements give in feet the actual encroachments since the 6-inch survey at definite points, from which the average encroachments on the coast line in the different 6-inch Yorkshire sheets mentioned have been deduced, viz.—

				Erosion.
Point A. Sheet 197 (Hornsea)	{ (Coast line surveyed but not examined yet)		.	165 ft.
,, B. ,,		ditto	.	198 ,,
C.		ditto	.	198 ,,
,, D. ,,		ditto	.	165 ,,

Average erosion in Sheet 197 = 182 ft.

			Erosion.
Point A. Sheet 213 (Aldborough) (Coast line examined)		.	244 ft.
,, B.	ditto	.	251 ,,
C.	ditto.	.	218 ,,
D.	ditto	.	244 ,,
E.	ditto	.	231 ,,
F.	ditto	.	231 ,,

Average erosion in Sheet 213 = 237 ft.

		Erosion.
Point A. Sheet 228 (Hilston)	. . . . .	132 ft.
,, B. ditto	. . . . .	165 ,,
, C. ditto	. . . . .	165 ,,
, D. ditto	. . . . .	99 ,,
, E. ditto	. . . . .	106 ,,
, F. ditto	. . . . .	158 ,,
, G. ditto	. . . . .	144 ,,

Average erosion in Sheet 228 = 139 ft.

		Erosion.
Point A. Sheet 243 (Withernsea)	. . . .	158 ft.
,, B. ditto	. . . .	207 ,,
C. ditto	. . . .	383 ,,
D. ditto	. . . .	198 ,,
E. ditto	. . . .	218 ,,
, F. ditto	. . . .	251 ,,
, G. ditto	. . . .	264 ,,
, H, ditto	. . . .	277 ,,
, H. ditto	. . . .	277 ,,

Average erosion in Sheet 243 = 245 ft.

Point A. North half of Sheet 257 (Holmpton) . . . 251 ft.
,, B. ditto . . . 251 ,,
, C. ditto . . . 244 ,,
: D. ditto . . . 277 ,,
E. ditto . . . 343 ,,
Average erosion in Sheet 257 N.=273 ft.

.Average erosion from Sheet 197 to 257 N. = 215 feet.

Average erosion per year = 5 feet 10 inches nearly ;
the years being 1889-1852 = 37 years.

Below is a list of measurements to the top of cliff (in
1895) from certain ancient and permanent fixed points
inland.

6-inch Sheet and Plan	Parish.	Name of Fixed Object.	Nearest Measurements in Feet to Top of Cliff.	Measurements due East in Feet to Top of Cliff.
197-3	Hornsea .	St. Nicholas' Church .	2,810.0	3,225.0
,,	,,	Do. chancel end . .	2,695.0	3,095.0
213-9	Aldborough	St. Bartholomew's Church	5,703.0	6,823.0
,,	,,	Do. chancel end . .	5,604.0	6,717.0
,,	,,	Old Windmill (Cen.) (Corn.)	5,307.0	6,364.0
228-7	Hilston .	St. Margaret's Church .	3,215.0	3,940.0
,,	,,	Do. chancel end . .	3,170.0	3,885.0
228-11	Tunstall .	All Saints' Church .	2,135.0	2,624.0
,,	,,	Do. chancel end . .	2,075.0	2,548.0
243-5	Withernsea	St. Nicholas' Church .	920.0	1,282.0
,,	,, .	Do. chancel end .	840.0	1,186.0
257-2	Holmpton .	St. Nicholas' Church .	3,203.0	4,093.0
,,	,,	Do. chancel end .	3,145.0	4,038.0

By the courtesy of the Director of the Ordnance
Survey I have been permitted to examine these maps,
and take many interesting particulars from them.

In *The Naturalist* for October 1904, Mr. R. G. Allanson-Winn gave an account of the sea's encroachment on
the east coast, and by the aid of diagrams endeavoured

6

to shew that in addition to the ordinary action of the sea

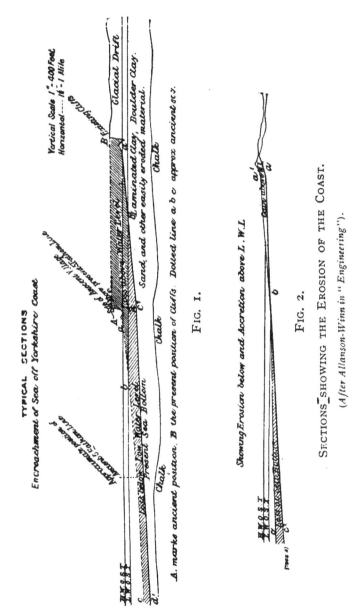

TYPICAL SECTIONS

Encroachment of Sea off Yorkshire Coast

Vertical Scale 1″= 400 Feet
Horizontal──── 1¼″: 1 Mile

Glacial Drift

Boulder Clay.

Laminated Clay, Boulder Clay, Sand, and other easily eroded material.

Chalk

Chalk

Chalk

Present Sea Bottom

A. marks ancient position. B the present position of cliffs. Dotted line a b c approx ancient sea.

FIG. 1.

Showing Erosion below and Accretion above L.W.L

FIG. 2.

SECTIONS SHOWING THE EROSION OF THE COAST.

(After Allanson-Winn in "Engineering").

ın wearing away the cliffs, there is a " Deep sea erosion "
taking place. " On the Holderness coast of Yorkshire

may be to-day pointed out situations, in the neighbourhood of the present five-fathom line, where formerly
towns and villages stood on the dry cliffs ; here we
observe the advance of deep water, and since it is im-
probable that the general inclinations of the shore and

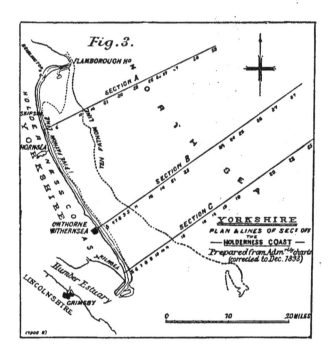

FIG. 3.

MAP SHOWING THE LINES OF SECTIONS ON PAGE 9.

sea-bottom have very materially altered since the old
days, we may fairly suppose that when these ancient
towns existed, the five-fathom line was a mile or so out
to sea, that is, two miles out from the present coastline.
In all cases, where the material is of a soft and easily-
eroded nature for a considerable depth below low-water

8

level, we have to consider the following :—(1) surf and

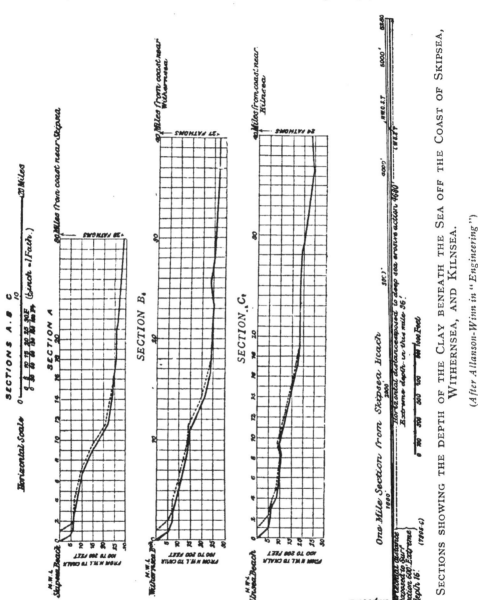

SECTIONS SHOWING THE DEPTH OF THE CLAY BENEATH THE SEA OFF THE COAST OF SKIPSEA, WITHERNSEA, AND KILNSEA.

(After Allanson-Winn in "Engineering")

wave action on visible shore between high and low-water ; (2) the erosion going on below low-water level

9

many miles out to sea and in 5 to 10 fathoms ; (3) the action of countless borers, worms, eels, shell-fish, etc., etc. ; (4) the action of submarine springs." The first two are considered to be the chief causes ; the latter two contributory causes only.

CHAPTER II

GENERALLY speaking, the English counties can be divided
into (*a*) those with the suffix " shire," as Yorkshire,
Lincolnshire, Hampshire ; and (*b*) those without, as
Surrey, Essex, Sussex. Historically, the ending " shire "
(which is derived from the Anglo-Saxon *scir*) is of considerable significance, as it indicates that it was originally
a share or division which made up one of the early Saxon
kingdoms. Thus, whilst Surrey, Sussex, etc., have survived from the old English kingdoms ; Yorkshire, that
is the shire of York, was once a part of Northumbria ;
and Hampshire, of Wessex. These " shares " were placed
under the charge of the chief, or ealdorman, and were
within easy access of a strong military centre. For instance, in our county the centre was York, which, even
in pre-historic times was a place of considerable influence.
While the divisions of England are generally referred
to as counties, not all of them are strictly so. Some, such
as Sussex, Essex, and Kent, reveal the Saxon origin of
their names.

Usually speaking, the boundaries of the counties are

fairly constant. There are, however, many factors which change these, notwithstanding that very largely the dividing lines are defined by natural features. The courses of rivers have altered, and in other ways changes have taken place.

Yorkshire is unique among the English counties in being divided into Ridings, namely, the East, North and West Ridings. The first-named is much the smallest, being barely a fifth of the entire area of Yorkshire.

The word " riding " is a corruption of " thriding," the old English form of " thirding," and is now used exclusively in this connection. The loss of the " th " is due to the mis-division of the compound word, as " North-thriding," " East-thriding," etc.

The East Riding is further sub-divided into Wapentakes, namely, the Dickering, Holderness, Hart-Hill, Buckrose, Howdenshire, and the Ouse and Derwent Wapentakes. The word " Wapentake " seems to have distinct reference to the military side of the organisation, and is generally connected with the Danish occupation. The Wapentake is only found in the Anglian districts of the country.

The East Riding being largely Wold country, and unsuitable for anything but farming, does not boast of many large towns, the greater part of the district being covered by more or less isolated farmsteads and small villages, with here and there a market town. Amongst the last-named are Malton, Driffield, Old Bridlington, Market Weighton, and Beverley, at which places the

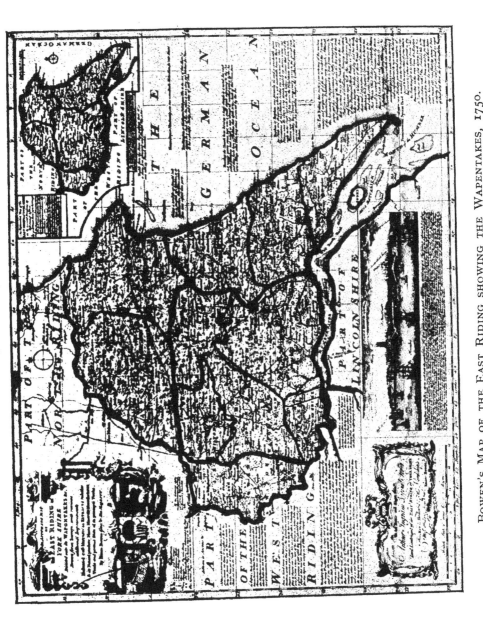

BOWEN'S MAP OF THE EAST RIDING SHOWING THE WAPENTAKES, 1750.
(It will be noticed that "Sunk Island" was really an island at that date).

farmers meet and carry on their business in the old-fashioned way. Generally speaking, all these are not as important to-day as they were years ago. The only place of really large size, which has recently received the dignity of being called a city, is Kingston-upon-Hull, and it occupies a position in about the centre of the southern boundary of the Riding. The one river of any importance, and that a comparatively small one, is the Hull, which rises a little west of Driffield, and flows almost due south to the Humber estuary.

The Wold country is high and dry, with a light chalk sub-soil ; while to the east, in the district of Holderness, the land is largely low-lying, and, as the sub-stratum is principally clay, it is difficult to farm. Parts of the Riding are very low indeed, and somewhat resemble the Fen-lands.

On the shores of the estuary, land has been reclaimed in several places. On account of their flatness, the Humber banks are somewhat uninteresting, though from an agricultural point of view, the land is all that can be desired.

The Holderness coast has quite a character of its own, consisting for the most part of low cliffs of Red or Purple Boulder-clay which is easily eroded. At its southern extremity is the long, narrow tongue of land known as Spurn Point, which is annually growing in a south-westerly direction, and is almost covered at the highest tides. A striking contrast with this is towards the northern extremity of the Riding's sea-board, where, at Flamborough Headland, the chalk cliffs rise in a sheer wall to a height of nearly 450 feet. Beyond this again is the dangerous ledge of rock known as Filey Brig, which juts out seawards almost at right-angles to the cliff.

The East Riding is somewhat pear-shaped, or perhaps can be better described as being the shape of a good Yorkshire ham, the " shank " being in the south-eastern extremity. It occupies about a fifth of the county, its actual area being 768,419 acres, though this figure undoubtedly changes annually, partly by reason of the great extent to which the east coast is washed away by the sea, and partly by the amount of new land which is reclaimed on its southern border.

The Riding is almost entirely defined by natural features; the east, from Filey to Spurn, being bordered by the North Sea; the south by the estuary of the Humber and the River Ouse as far as York; the north and north-west almost entirely by the River Derwent, the source of the latter being near to the coast.

Compared with the West and North Ridings, the East Riding has no pronounced physiographical features. There are no mountains; the hills are comparatively small, and are not broken up into isolated masses, the result being that it has a very regular, if undulating, aspect.

The Riding is divided into three very distinct areas, namely, the Plain of York, the Wolds, and Holderness.

The first of these includes the whole of the land to the north and west of the Wolds, and consists of about

200,000 acres. The western portion is low and flat, and, whilst the Triassic and other rocks occur beneath, they are covered with silts, blown sands, and other newer deposits. In this area are tracts of sandy moors, which in places still retain their original character.

The Wolds occupy the centre of the Riding, and form its backbone. At their highest point they do not quite reach 800 feet above sea level. They occupy about 400,000 acres, and curve round in a broad belt from the north-eastern part of the Riding to its southern-most extremity at Hessle, and, with the exception of a few isolated places of very small extent, are not covered by any superficial deposit. The soil is light, with a large admixture of chalk, and, in places, flint. There is a small gap in the Wolds at Market Weighton, but otherwise they have rounded contours very much after the style of the Downs in the south of England. On the west there is a more or less pronounced cliff, or escarpment, which is flanked by the blown sands of the Vale of York. On the east, however, the slope is more gradual owing to a plastering of drift, principally boulder-clay, which was effected during the Great Ice Age.

The Wolds are widest in the north, and gradually narrow towards Hessle, where they have been cut through by the Humber ; the range of hills being clearly seen on the opposite side of the estuary near Barton in Lincoln-shire.

Between the Wolds and the North Sea lies the Plain of Holderness, with an area of 160,000 acres. This distriet is saucer-shaped, being highest on the east, west, and south, the centre being occupied by the River Hull and its connecting water-courses. The sea-board of Holderness is about 35 miles in length, and the cliffs on an average are not more than 30 feet in height. At one point, towards the south, at Dimlington, they are about 150 feet high, and are entirely composed of glacial clays.

At one time most of Holderness was under water, and would then greatly resemble the Norfolk Broads. Of these numerous meres, but one remains, that at Hornsea, which is 1½ miles long and half a mile broad. The sites of others are clearly defined, and some of them are exposed in section in the cliffs. The more important of these are at Bridlington, Barmston, Skipsea, Atwick, Hornsea, Sand-le-mere, Withernsea, Out Newton, and Kilnsea. Inland the positions of the meres are indicated by the place names, such as Marton, Marfleet, Rowmere, Giltsmere, Redmere, Braemere, and the Marrs.

The surface of the district, though usually low-lying, is by no means flat, except in the valley of the River Hull. This is due to the fact that the whole area is composed of glacial morainic material, which is hummocky; the hills and hollows occurring without any apparent order, beyond that, generally speaking, there is a line of gravel mounds extending in a southerly direction from Bridlington, through Kelk and Brandesburton, towards Paull on the Humber. These mounds are usually composed of clay, with cores of gravel, and occasionally, in the northern part of the district, they are entirely made up of fine sand and gravel.

The only river worthy of the name, which is exclusively in the Riding, is the Hull, which rises from a spring in the chalk at Emswell, near Driffield, and flows almost due south to the Humber, where the city of Kingston-upon-Hull occupies its banks. Formerly the river was fairly wide and sluggish, and was undoubtedly tidal for the greater part of its course. To-day it is confined to narrow limits by embankments, which at its southern extremity are yet occasionally being built further into the river. The stream is now tidal only as far as Beverley, where the upper reaches are cut off by lock-gates. The river, however, is navigable for sloops and other small craft as far as Driffield.

The Hull is fed on both sides by a few small natural streams, known as becks, which are attractive to anglers, and by a great number of artificial drains.

The only other water-courses in the Riding are exceedingly small and insignificant, and reach the sea or Humber at Speeton, Bridlington, Barmston, Out Newton, Hessle, and near North and South Cave respectively. To the west also is the River Foulness,* which now joins the Market Weighton Canal.

Bordering the Riding for almost its entire northern and western boundary, however, are the Rivers Derwent and Ouse. The Derwent is the main left tributary of the Ouse, and is connected with the parent stream at Barmby-

* Pronounced "Foonay."

on-the-Marsh, having thus for a part of its course been entirely within the Riding.

Such watershed as can be said to exist is unquestionably the ridge known as the Wolds, which approximately divides the Riding into two parts. The Wold area, however, is remarkable for its comparatively great breadth without a single stream. On the west a few small watercourses join the Derwent, while on the east is the Hull and its tributaries.

The Wolds, being fairly flat, and exceedingly porous, easily absorb rain, and this, owing to the dip of the rocks, travels in a southerly direction. For this reason the supply of fresh water at Hull is almost inexhaustible. At Mill Dam and at Springhead are large pumping stations for the Hull Corporation. The North Eastern Railway Company also has a pumping station at Hessle, and in addition, great quantities of fresh water run to waste at " Hessle Whelps," on the north shore of the Humber.

Not connected with the watershed in any way, but merely resting in a hollow in the Drift, is Hornsea Mere, the largest sheet of fresh-water in Yorkshire. The raised gravel terraces on its sides indicate that it was once considerably larger in extent. It is now very shallow, and in the summer is resorted to by owners of small pleasure yachts, and also to a large extent by anglers. In severe winters it affords excellent skating.

CHAPTER III

DIVISIONS OF THE EARTH'S CRUST—THE YORKSHIRE
CLIFFS—LIAS, OOLITES, CHALK—THE GREAT ICE
AGE—FORMATION OF HOLDERNESS—STRUCTURE OF
HOLDERNESS.

IN England we are exceptionally favoured by having
representatives of most of the important strata comprising the earth's crust, and in our islands we have an
epitome of the geological history of the earth. In
Yorkshire are examples of most of the important rocks
of the country.

Of the four great divisions of the earth's crust which
have been made for convenience of classification, there are
represented in East Yorkshire the beds of the Secondary
or Mesozoic rocks, and the Quaternary or recent rocks,
only ; but of the oldest or Palæozoic, and of the great
Tertiary system, which is so admirably developed in the
south-east of England and in France, we have practically
no trace whatever. There is no doubt that at some time
East Yorkshire was covered by some of the Tertiary
beds, but they have since been entirely swept away.

Confining ourselves to the beds of sandstone, limestone, and shale formed during the Secondary epoch,
and to the sand, gravel, and clay deposited in comparatively recent times, we find that they reveal many

20

interesting facts in relation to this part of the county. In East Yorkshire it may be truly said :—

> The hills are shadows, and they flow
> From form to form, and nothing stands ;
> They melt like mists, the solid lands,
> Like clouds they shape themselves and go.

GRAVEL CARRIED DOWN THE VALLEY AS A RESULT OF A CLOUDBURST AT COWLAM, 1910.

The deposits within easy reach of Hull give absolute evidence of many changes having taken place. We can see these occurring to-day on a small scale. As shewn in another chapter, our cliffs for 30 miles are being eroded at an average rate of 7 feet per annum. Our rivers are constantly carrying detritus into the Humber and the sea. The rain, frost, and wind are slowly but surely affecting our wolds, dales and cliffs. Occasionally a water-spout or other unusual phenomenon reminds us of the great power of the " elements " when they have full play. A large gully in the side of the dale at Langtoft, near Driffield, is evidence of one of these ; and only quite recently a similar rift was made in the Wolds near Driffield by a cloud-burst. Such changes, though apparently small and insignificant, unquestionably tell in time.

EFFECT OF CLOUD-BURST AT COWLAM, 1910.

Compared with geological time, our centuries are but seconds. The formation and erosion and re-formation of the various rocks, time after time, as well as the gradual evolution of animal and plant-life, the remains of which are therein entombed, require a period which, if numbered in years or even centuries, would be far too great to be comprehended. In the geological history of East York-

22

shire we are starting at a point very late in the history of the globe. The remains of animals and plants preserved in the Liassic strata, our oldest fossiliferous beds, indicate a highly advanced state. The relative position of the Liassic rocks upon the older series shews that they were formed at a comparatively recent period, and great as are the changes which have taken place between then and now, these are small and insignificant compared with the upheavals and depressions which had taken place prior to the formation of the Liassic beds.

The rocks of East Yorkshire are exposed in the cliffs in a remarkably regular order. The beds are deposited one upon another, just as one may pile up some volumes of books, and there is a gradual dip or slope of the rocks to the 'south-east. Thus, during a walk along the cliffs from south to north, it is possible to examine each bed in its proper order, as it presents its surface in the cliff line.

The Lias, as the name implies, is that division of the rocks characterised by alternate thin *layers* of limestone and shale. In East Yorkshire the beds can be examined at North Cave and Cliff ;- at which places are some lower beds in the series than are represented in the well-known coast sections near Whitby.

After the deposition of several hundred feet of these beds in the Liassic sea, the area was elevated, and the water and its inhabitants found a resting-place elsewhere. What was once an ocean floor became for the time being dry land, which eventually again found its level beneath the sea ; and upon the Liassic rocks was deposited a great series of limestones and shales of newer date, belonging to the Oolitic system. These various beds are shewn in the table of strata on page 35.

Space does not permit us to examine each of these in detail, but they comprise a series of rocks altogether

A "SNAKESTONE."

measuring some thousands of feet in thickness. They indicate many changes in the earth's crust during their formation. Take for example the Millepore Bed, so-called from a minute coral with numerous small pores occurring therein. It is a characteristic roe-stone or Oolitic rock. If a specimen, say from Hotham or Brough, be carefully examined, it will be found to consist of minute globules of lime, cemented together, very much resembling the roe of a fish. In the matrix of this rock are innumerable fragments of the small coral already referred to, as well as remains of ammonites, and various bivalve and univalve shells, characteristic of the series.

The Kellaways rock, another Oolitic bed, comes next, and is evidently a shallow-water formation. At South

Photo by]   FILEY BRIG FROM THE NORTH.   *[Godfrey Bingley.*

Cave an excellent section is exposed in the railway cutting. It abounds in various molluscan remains, and at its base is a well-defined oyster-bed, 8 inches in thickness, crowded with thousands of oysters.

Professor Kendall has shewn that while great masses of rock were in process of formation in some parts of East Yorkshire, there seems to have been some disturbing element in the Market Weighton district, which interfered with the regular deposition of many of the beds in that particular area. It is clear that during the long Secondary period there was a gradual lifting of the earth's crust in the Market Weighton area, which resulted in only thin beds of the various rocks being deposited there.

On leaving the Oolitic rocks, with their coral and

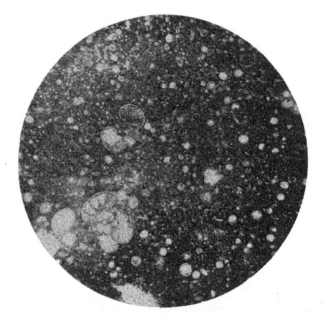

SECTION OF CHALK, MAGNIFIED.

shore deposits and shallow-water formations, we find evidence of a change of a somewhat drastic character. The land here, in common with the greater part of Europe, seems to have been depressed. It was under deep water, and was slowly but surely covered by a soft white sediment now known as chalk. In parts of Britain, and particularly in East Yorkshire, traces of this chalk sea still remain, but over large areas

26

every fragment of it seems to have been swept from the face of the land. Mr. G. W. Lamplugh has estimated the thickness of the Yorkshire chalk to be 1270 feet. If we take a piece of this chalk, say from Hessle or Flamborough, and, after preparation, place it under the microscope, it will be found to consist of the minute skeletons of small organisms known as Foraminifera, almost precisely similar forms to those which occur at the bottoms of our deep oceans to-day. In almost any chalk-quarry in the Riding one may find teeth of sharks, occasional remains of fishes, tests of sea-urchins, large oyster-like bivalves, remains of ammonites, cuttle-fishes, sponges, and other denizens of the deep.

This area was subsequently subjected to another change ; the old ocean floor being elevated to a great height above the level of the water. The hundreds of feet of solidified ooze were thus raised and formed an enormous continental plateau. Probably at this period originated some of our present river-channels.

There is no doubt that many of the sites of our present rivers are of great antiquity.

Between the close of the Cretaceous period and the next deposit in East Yorkshire is a great gap in our geological history. There is not in Yorkshire, so far as is at present known, a vestige of that great series of marls and gravels forming the Tertiary system, which, as already pointed out, occur in great thickness elsewhere.

What was taking place in Yorkshire during this enormous period of time it is difficult to say.*

The next stage in our geological history is marked by a channel formed in the chalk plateau by an immense river running along in a northerly direction, in the centre of what is now the North Sea. Of this river the Thames, Humber, and some of the present Continental streams, were tributaries. Eventually the ocean carried on the work started by the river, and between what is now the Continent and Britain was a gradually widening arm of the sea.

On the west was a long line of chalk cliffs, which gradually receded towards our present shores, until finally a white wall, a hundred feet in height, presented its front to the sea, which was then washing over what is now Holderness. This cliff still lies buried beneath the glacial drift. At Sewerby, near Bridlington, the junction of the old and new cliff can be seen ; from which place it curves round through Driffield, Beverley and Hessle, where it has been exposed during quarrying operations. Borings for wells show that buried under Holderness is a shelving rocky floor, sloping from the old cliff foot towards the east.

At this stage the sea seems to have been arrested in

* In many places on the chalk-wolds are " pipes," or deep holes in the rock, generally only a foot or two in width, but extending to a great depth. These contain numbers of well-rounded pebbles of quartzite. Similar pebbles also occur below the glacial beds at Hessle and Sewerby. As they have certainly not been derived from the Drift, they can therefore only represent some deposit which at one time must have covered the chalk ; but which, previous to the Glacial period, was entirely swept away.

28

its work. The Ice Age, and the débris then dumped down in this district, entirely altered the configuration of the country. On the melting of the ice, however, the sea renewed its energies, and at the present time is gradually approaching its long-neglected cliff line.

Banked up against the perpendicular face of the old cliff are gravels and sands representing ancient beach deposits, covered by a quantity of angular chalk " wash " and blown sand. All these were formed before the advance of the ice, as the Drift covers them.

A careful examination of this pre-glacial beach shows that during its formation some change was going on in the district ; the water seems to have been receding from the cliffs, and the animal and plant remains indicate colder conditions. In other parts of England the Tertiary beds clearly prove that prior to the Glacial Period the climate of the country had been gradually cooling.

It is evident that small glaciers began to form in the mountainous districts of Scotland, Wales, the Lake District, and Scandinavia. These ice-streams increased in size and eventually coalesced and flowed into the seas. From the English Lake District a huge glacier 20 miles wide filled Teesdale, and entered the North Sea on the site of the Tees' present mouth. An arm of this glacier broke through the valley side, and descended into the Vale of York, where, near York and Tadcaster, it left two of the most perfect terminal moraines to be found in these islands.

About the same time the Scandinavian ice entirely

occupied the bed of the North Sea, and on reaching our shores, its force was such that it diverted southward the enormous Teesdale glacier. The combined ice of Teesdale and the more northern streams coasted along East Yorkshire from north to south ; the cliffs of Speeton, 440 feet in height, formed too abrupt a buttress to be entirely surmounted, and a great moraine, consisting of gravel and clay, was deposited along the cliff edge, between Speeton and Buckton. Speeton windmill stands on one of these morainic mounds. On reaching the lower part of the headland, however, where the chalk is only about 200 feet in height, the ice over-rode the land, and a thick bed of boulder-clay was left when the ice melted.

Rounding the corner, the old Bay of Holderness was entirely occupied by the invader, up to, and beyond the pre-glacial cliff ; and the great mass of gravel, clay and sand, forming Yorkshire, east of the Wolds, is merely the moraine left by the ice.

In the gravel-pit at Burstwick the nature of this material can be seen. Rocks from Scotland, the Cheviots, and the English Lake District, occur cheek by jowl with the limestones of Teesdale, fossils from Whitby, Scarborough and Flamborough, and rocks from Christiania. In addition are innumerable fragments and perfect examples of shells of an Arctic type, torn from the bed of the North Sea by the glacier, and with these are bones of walrus, elk, elephant, and other animals.

There is a line of gravel mounds extending from

THE NEW KELSEY HILL GRAVEL PIT, A SECTION IN THE HOLDERNESS MORAINE.

Bridlington through Kelk, Brandesburton, Sproatley, and Burstwick towards Paull, which represents material deposited by the melting ice at one stage of its career. The mound at Paull is no doubt responsible for the diversion in a south-westerly direction of the present Humber channel. Originally, as might be expected, and as borings in Holderness prove, the Humber flowed due east, and the mantle of Drift which made so many changes in the Yorkshire river channels caused the Humber to change its course. Grimsby would not now be a port were it not for the mound of gravel at Paull which was deposited during the glacial period!

CHAPTER IV

PHYSIOGRAPHICAL CHANGES—GLACIAL LAKES—CHANGES
IN RIVER COURSES—TABLE OF STRATA—PEREGRINA-
TION OF THE COAST—HEADLANDS AND BAYS—GAINS
AND LOSSES ON THE COAST—HISTORICAL EVIDENCE—
PLANS AND MEASUREMENTS.

FORMERLY the River Derwent emptied itself due
eastwards into the sea. Its outlet, however, was blocked
by the Drift, its waters were diverted into an entirely
different direction, and to-day have to flow round the
East Riding by way of Helmsley and Goole and the
Humber.

During the Great Ice Age the damming of the entrance
of the Humber by the glacier, formed " Lake Humber,"
which must have been a sheet of water of truly gigantic
proportions, as can be readily understood if one assumes
for the moment the probable effect of damming the
waters of the estuary to-day. In glacial times, there
were two barriers across the Humber; the one already
mentioned at Paull, and one crossing from North to
South Ferriby. Remains of the latter still occur on the
river banks.

On the final melting of the ice, East Yorkshire would
present a dreary aspect. Huge fans of gravel were
formed at Bridlington and other places as a result of the
floods. Rounded hillocks of Drift stood out in all

C

directions, the hollows between being filled with miniature lakes or meres. Of animal or plant life there was
little or none. As the climate became milder, small
numbers of Arctic animals and plants made their appearance. Later the margins of the meres became clothed
in vegetation, peat was formed, and huge oaks and firs

Photo by]   [*J. Hollingworth.*
SECTION IN THE CLIFF SOUTH OF BRIDLINGTON,
NOW HIDDEN BY THE NEW SPA.

The section shows "late-glacial" gravels formed at the close of the
Great Ice Age.

flourished. The red deer and other animals haunted the
woods, and, after a time, man made his appearance, and
lived on the holmes or islands, or on pile dwellings on
the edges of the lakes. Largely as the result of natural
and artificial draining, these conditions have been changed,

34

and there is hardly any land but what is under cultivation to-day.

GEOLOGICAL FORMATIONS OCCURRING IN THE EAST RIDING OF YORKSHIRE.

(From H.M. Geological Survey Memoirs).

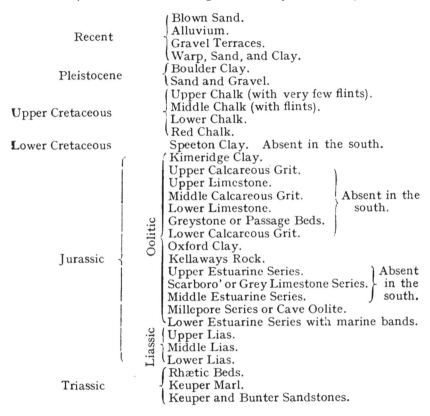

Recent	Blown Sand.
	Alluvium.
	Gravel Terraces.
	Warp, Sand, and Clay.
Pleistocene	Boulder Clay.
	Sand and Gravel.
Upper Cretaceous	Upper Chalk (with very few flints).
	Middle Chalk (with flints).
	Lower Chalk.
	Red Chalk.
Lower Cretaceous	Speeton Clay. Absent in the south.
Jurassic — Oolitic	Kimeridge Clay.
	Upper Calcareous Grit.
	Upper Limestone.
	Middle Calcareous Grit. } Absent in the south.
	Lower Limestone.
	Greystone or Passage Beds.
	Lower Calcareous Grit.
	Oxford Clay.
	Kellaways Rock.
	Upper Estuarine Series.
	Scarboro' or Grey Limestone Series. } Absent in the south.
	Middle Estuarine Series.
	Millepore Series or Cave Oolite.
	Lower Estuarine Series with marine bands.
Jurassic — Liassic	Upper Lias.
	Middle Lias.
	Lower Lias.
Triassic	Rhætic Beds.
	Keuper Marl.
	Keuper and Bunter Sandstones.

Commencing at Filey, where the East Riding joins the North, we find a low perpendicular wall of Oolitic limestone forming the base of the cliffs; the upper part consisting of soft boulder clay. This latter material is only slightly affected by the sea, but is carved into steep ridges and pinnacles by sub-aerial denudation. The limestone cliffs are weathered into numerous miniature bays, locally known as " doodles." A particularly solid bed of rock resists the action of the sea and forms Filey Brig, adjoining the south side of which is the " Spital," said to be of Roman construction. Doubtless at one time this ledge of rock was Drift covered.

Photo by]   FILEY BRIG.   *[Godfrey Bingley.*

Between the Brig and the next promontory, Flamborough Headland, is Filey Bay, which owes its existence to the comparative ease with which the Drift and the soft Oolitic and Neocomian clays have been denuded. The Drift in Filey Bay no doubt hides a Pre-Glacial river-bed.

Flamborough Headland, the most striking feature on the east coast of England, is a hard mass of chalk, the north side of which is an almost straight, perpendicular wall, reaching 440 feet in height, with a very thin covering of Drift on the top. In places the sea never recedes from

BOULDER CLAY CLIFFS AT WITHERNSEA.

the cliff foot, and there is deep water quite close to the land.

South of Flamborough the chalk cliffs are softer and comparatively low, but have a thick capping of Drift,

37

which, as at Filey, weathers into picturesque pinnacles and ridges.

This part of the Headland is riddled with innumerable small bays and caves. The roofs of some of the caverns near the lighthouse have fallen in, and form the " blow-holes," through which water and spray is driven in severe weather.

From Bridlington to Kilnsea is the Bay of Holderness,

Photo by] THE CLIFFS AT KILNSEA. [*C. W. Mason.*
Showing the way in which the sea is assisted in its work of destruction by the drains leading to the cliff edge and softening the clay.

with its low boulder-clay cliffs, at the foot of which is an excellent sandy beach. At its south extremity is the long narrow bank of sand-dunes known as Spurn Point, held together partly by natural vegetation, partly by groynes and other artificial contrivances. There is a lighthouse at the Point, and another a few miles to the

38

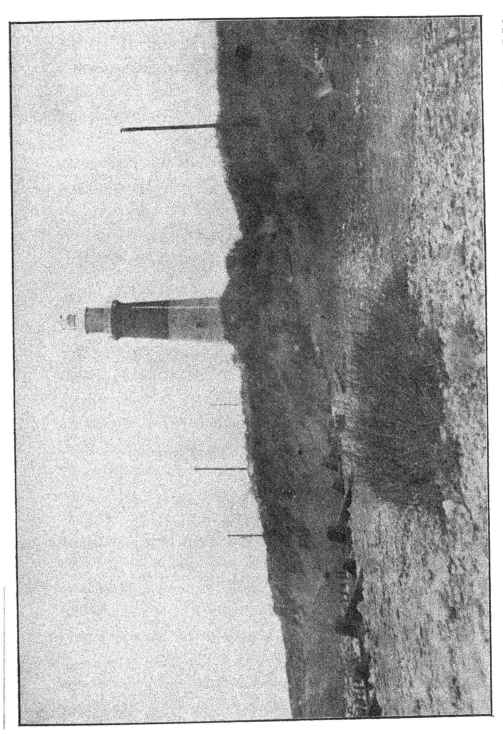

SPURN LIGHTHOUSE.

north-west, at Withernsea. From Spurn, along the
Humber Estuary, the land is low and flat, and protected
for the greater part of its length by earthen embankments. At Paull, east of Hull, is a large gravel mound
on which the battery is built, and at Hessle and Brough
are the southern extremities of the Wolds, though even
here a stretch of flat silty or gravelly land adjoins
the Humber. Towards the western extremity of the
southern portion of the Riding the land is also
marshy or waste, and has to be protected by embankments.

Probably no district in the British Isles offers such a
variety of lessons as does that between Bridlington and
Spurn Point, and the Humber Estuary. On the one
hand enormous tracts of land have disappeared within
historic times, while on the other, large areas have been
formed, embanked, and cultivated. This district is
also especially worthy of attention from the fact that,
relating to these changes, data of an exceptionally
reliable character are available.

At the close of the Great Ice Age, East Yorkshire
would begin to assume something of the form familiar to
us to-day. The coast-line, however, was some miles to
the eastward of its present position. Spurn Head was
not formed; Sunk Island, Broomfleet Island, Read's
Island, and other reclaimed lands in the Humber, were
not then known; certainly large tracts of country, such
as the Valley of the Hull, were tidal, and only became
habitable by a process of natural silting up, which is now

40

artificially carried on in some parts of the district by warping.

In three ways we have valuable historical evidence of geographical changes :—(1) that of Domesday and other documents ; (2) that supplied by old maps and plans ; (3) that obtained by comparing old measurements from various points to the cliff edge, with modern measurements from the same points.

(1) In his paper on " The Ploughland and the Plough," the late Canon Isaac Taylor showed that in Domesday Book, two and three-field manors are distinguishable by the relative number of carucates and ploughs. In three-field manors the average number of ploughs is equal to half the number of carucates, whilst in two-field manors the number of ploughs and carucates, as a rule, is the same. Canon Taylor also showed that in Yorkshire at least the carucate is a definite measure of arable land. The Domesday measurement formed the basis of taxation, and, as in a two-field manor, in any year only half the arable land was in culture, it follows that the Domesday measurement in such cases represents only one half of the actual arable area. In three-field manors, on the other hand, as only one field lay fallow, the Domesday measurements represent two-thirds of the whole arable area. Canon Taylor has further shown that the amount of arable land in a Yorkshire manor, at the time of the survey, was practically identical with the extent of such land at the period of enclosure.

It may be fairly assumed (a) that in a number of

townships the average relation between the area of arable land would not greatly vary, and therefore (*b*) that the loss of arable land in the coast townships of Holderness will indicate to a fairly accurate extent the loss of other lands.

In this way it is possible to compare the extent of arable land in four coast townships in East Yorkshire in the year 1086 and, say, in the year 1800 :—

	1086.	1800.
Easington	2400 acres	1300 acres
Holmpton	1280 ,,	900 ,,
Tunstall	1280 ,,	800 ,,
Colden	1920 ,,	1100 ,,

The coast townships of Holderness now contain 33,468 acres. Assuming that the waste of other kinds of land has been equal to the arable land in the other townships, the sea has carried away 22,694 acres, or over $35\frac{1}{2}$ square miles. The coast of Holderness, from Barmston to Kilnsea, is $34\frac{1}{2}$ miles long. Assuming the land has been carried away equally throughout the distance, there has been a strip of land 1809 yards, that is a little over a mile wide, washed away, at the average rate of 7 feet 1 inch per annum. If the denudation since Roman times has been at the same rate, 53,318 acres, or about 83 square miles, have been lost. This is equal to a strip of land $2\frac{1}{2}$ miles in width for the whole of the distance.

It is remarkable that this estimate precisely agrees with that ascertained in recent years by actual measurement.

From a gaudily coloured " Plotte made for the descripcion of the river Humber and of the Sea and Seacoost from Hull to Skarburgh," in the late years of the reign of Henry VIII., Hornsea is shewn at some distance from the sea, and is connected therewith by a long straight river or creek.

This and other old plans, referred to in detail later, show that right along the coast many places of importance, as well as smaller villages, have disappeared. Some, as at Kilnsea, Hornsea, etc., occupied positions to the east of the present townships bearing these names, the places gradually travelling westwards as their eastern portions were washed away. Others, such as Owthorne, Auburn, Hartburn, and Hornsea Burton, have been entirely washed away.

On Tuke's map of 1786, and in papers and books innumerable, are particulars of scores of measurements from fixed points near the coast-line to the cliff-edge. In recent years these various places have been re-measured by members of the Yorkshire Coast Erosion Committee and others. Without going into details, it may be stated that these measurements prove that the actual wastage of the thirty-five miles of cliff between Bridlington and Spurn is 7 feet per annum. Near Hornsea the average loss appears to be 4 feet per annum ; at Sand-le-mere, Withernsea, and Holmpton, 9 feet per annum ; and between Kilnsea and Spurn Point, 6 feet per annum. At various points along the coast, however, exceptionally heavy falls of cliff occasionally occur, which would considerably increase these figures if only a few years were taken into account.

CHAPTER V

RECLAIMED LAND—SUNK ISLAND—BROOMFLEET ISLAND
—READ'S ISLAND—LOST TOWNS OF THE HUMBER—
RAVENSER—HULL—HEDON.

As years go on the Humber is confined to narrower
and narrower channels. A few centuries ago large tracts
of land were under water. These, having been reclaimed
and embanked, are now producing good crops. Earlier
still, the Valley of the Hull and other large areas were
tidal arms of the Humber, and the town of Hull itself is
built on an accumulation of silt which was not existing
in Roman times.

Not only have these connecting valleys been silted up,
but great areas have been reclaimed ; some probably so
long ago as in Danish times ; others more recently. Of
comparatively late date are those large areas, Sunk
Island, Broomfleet Island, and Read's Island ; though the
last-named only, is now an island proper.

The growth of Sunk Island can be readily traced by
the various maps and charts of the Humber, which have
been published from time to time. In the earliest of these
a mere sandbank is shown ; then an island, which increases
in size until eventually the channel between it and
Holderness is silted up, and the whole is joined to the
mainland. So long ago as 1660 two large sandbanks

44

were shewn to the west of Paull, and the gradual growth of these has resulted in the addition of about 7500 acres of excellent farm land to this part of the country. In 1774, 1561 acres had been enclosed; 26 years later, Cherrycob Sands were enclosed; further reclamations were made in 1826 and 1850; and in 1880, 2700 more acres were reclaimed. More recently, in 1897, 347 acres

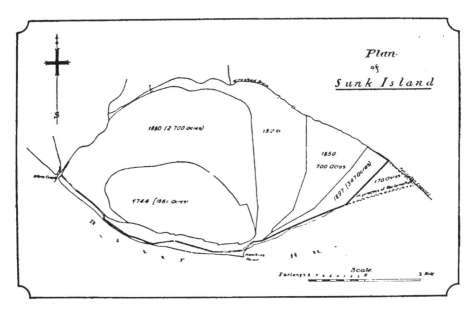

SUNK ISLAND, SHOWING THE VARIOUS AREAS RECLAIMED.
(*After Butterfield*).

were embanked, and a further 170 acres are now being reclaimed.

The following are the dates of making junction banks from the mainland to Sunk Island:—" 1772, the clough at No Man's Friend erected; 1799, Mr. Watt enclosed the growths of Ottringham by a bank at the west lands jetty, which . . . divides the townships of Pattrington

45

and Ottringham, and carried the Ottringham drainage
waters into Stone Creek; 1819-20, Col. Maister enclosed

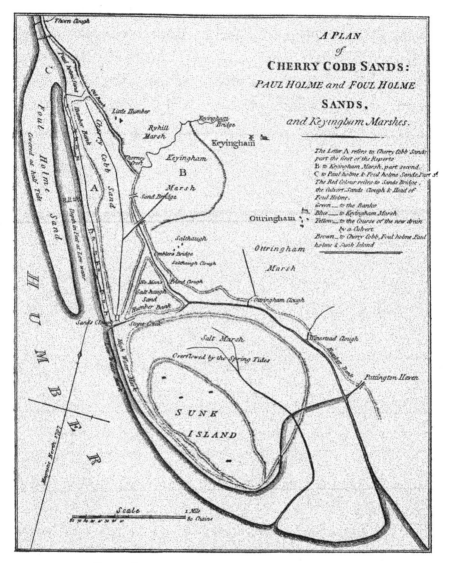

SUNK ISLAND AND SURROUNDINGS IN 1797.

the west land growths by a bank across the channel to
join Sunk Island, and make a communication with it;

46

1839, Pattrington west growths completely embanked, Winestead clough brought down the channel near Pattrington Haven mouth ; 22nd June, 1841, a good road made over it to Sunk Island."

A century ago, large accumulations of sand were formed on the north Humber · shore at Broomfleet Island, west of Brough. By 1820 they had reached

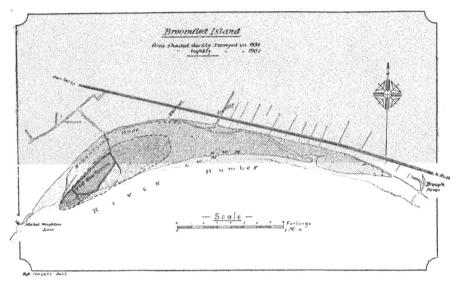

BROOMFLEET ISLAND, SHOWING NEW LAND RECLAIMED.

a high level above high-water mark. In 1846, James Oldham reported on enclosing 130 acres, but soon afterwards it began to disappear, and the whole island went, as well as part of the mainland. In 1853 it began to accumulate again, and in 1866 six acres were enclosed. In 1870 there were 60 acres embanked, the channel between the island and the mainland gradually silted up, and by 1900 was entirely closed. At the present day nearly 600 acres have been added to the county at this point.

Read's Island, still an island, is between Brough in Yorkshire and South Ferriby in Lincolnshire. The channels on either side, however, vary; in my time the shipping has had to change its course from one side to the other two or three times. The island

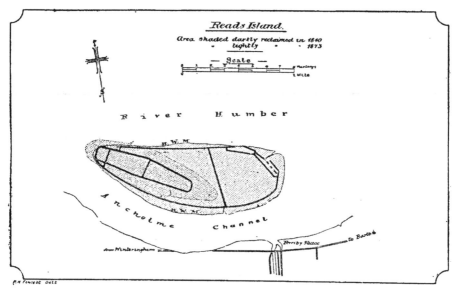

READ'S ISLAND, SHOWING THE VARIOUS AREAS RECLAIMED.

commenced to form early in the nineteenth century. In 1840, 75 acres were enclosed. In 1861, there were 289 acres embanked. In 1886, 491 acres, 450 of which were enclosed. Since that date the island seems to have ceased growing, and is now being washed away at an average rate of $4\frac{1}{2}$ acres a year; 67 acres having disappeared between 1886 and 1901.

Just as thousands of acres of new land have been

formed within the Humber estuary in comparatively recent times, so in the past we find villages, and even flourishing towns, have disappeared. The late J. R. Boyle, in his " Lost Towns of the Humber," brought forward unquestionable evidence of many important places having existed on sites which are now within the waters of the estuary, or have been recently covered up

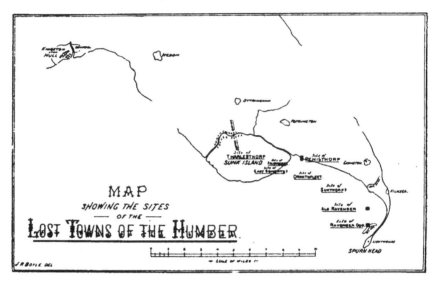

THE SITES OF THE LOST TOWNS OF THE HUMBER.
After the late J. R. Boyle (compare with frontispiece).

by the accumulation of mud-flats. Among the places which have disappeared are Tharlsthorp, Frismersk, East Somert, Penisthorp, Orwithfleet, Sunthorp, Ald Ravenser, and Ravenser Odd.

With regard to Ravenser, this place seems clearly to have originally been a Danish settlement. The first part of the word relates to the Danish standard, a raven ; and the second portion, *cyr* or *ore*, " denotes a narrow strip

of land between two waters." It is interesting to remember that Ravenser is referred to at least three times
in ancient Icelandic literature, in connection with the
battle of Stamford Bridge, references to which in the
Saxon Chronicle are well-known. The following passage
relating to the departure of the fleet from Ravenser is
of interest in this connection :—

" Olafr, son of Harold Sigurdson, led the fleet from
England, setting sail at *Hrafnseyri*, and in the autumn
came to Orkney. Of whom Stein Herdisson makes
mention :—

> " The king the swift ships with the flood
> Set out, with the autumn approaching,
> And sailed from the port called
> Hrafnseyri (the raven tongue of land).
> The boats passed over the broad track
> Of the long ships : the sea raging,
> The roaring tide was furious round the ships' sides.'

The story of Hull itself, though one of continued
growth and prosperity, also has its interest from the
physiographical changes which its old records indicate.
In the earliest times, what is now the River Hull, certainly
had two mouths, the " Old Town " itself being built upon
the delta between. The records also shew that in the
immediate vicinity of the town the land was of a marshy
character, and in many places absolutely impassable.
Advantage of this state of things was taken during the
Civil War in 1642, when, by a series of sluice-gates
connecting the Humber and Hull, the entire surrounding
country was flooded. The various plans of the town

and district, of which an exceptionally complete series

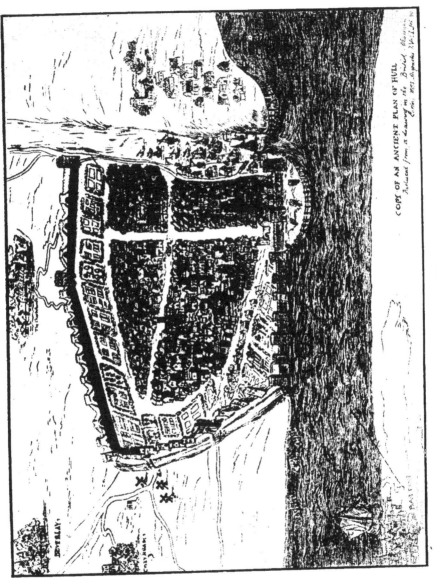

HULL IN THE XIV. CENTURY.

is available,* clearly .indicate the enormous changes which have gone on with the growth of the town.

* " The Evolution of Hull," pp. 203, 1911.

51

It has much changed since it was a small vill called Wyke. That was before the days of King Edward I. In mediæval times it was a walled and moated stronghold, and of some consequence in the Civil War. Early in the eighteenth century the old wall, moat and adjoining lands were handed over to the Hull Dock Company, and upon their site a line of docks was built. This accounts for the " Old Town " of Hull still being an island connected with the outside land by bridges. As time went on these docks were too small for the town's requirements, and new docks were made, and are still being made to the west and east of the town. These are along the Humber side, their construction extending the town's frontage southwards in consequence of retaining walls being built into the Humber. " Sand South End," where two centuries ago the obstreperous ladies of Hull were ducked into the Humber by means of a ducking-stool, is now well inside the " Old Town."

The once flourishing seaport of Hedon, which sent three members to Parliament, and still possesses valuable corporation plate, including the oldest mace in the country, is now a quiet country town of a thousand inhabitants. It is to-day some two miles distant from the Humber, though the old docks and waterways can still be traced in grass fields. Its official seal—a sailing ship manned—seems to be all there is bearing upon its one-time connection with the sea. A narrow meandering creek at high tide now allows small craft to approach a

View of the Old Garrison at the Mouth of the River Hull, *civ.* 1850, by J. Ward.

(The original in the Wilberforce Museum, Hull).

town which once supplied ships and men to the king's navy.

It will thus be seen that in some instances ports have disappeared simply from the fact that they have been entirely washed away by the sea, and consequently their trade has been diverted to other channels. In the case of Hedon, however, the silting up of its connecting channel with the Humber has had a similar effect.

The Western Haven, Hedon, silted up.

VIEW OF HULL FROM NEW HOLLAND FERRY HOUSE,
BY JAMES SHEPPARD, 1832.

The number of Windmills is remarkable.   The position of the
"Old Harbour" is shown by the number of ships' masts.

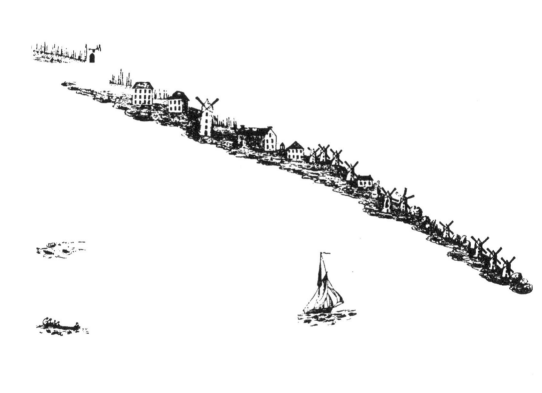

CHAPTER VI

THE DRAINAGE — HOLDERNESS *TEMP.* EDWARD I.—
ANCIENT EVIDENCES OF DRAINAGE—EIGHTEENTH
CENTURY DRAINAGE.

SEEING that a large part of Holderness is below sea-level, and we have geological evidence that the valley of the Hull and other districts were once accessible to the waters of the sea, and that in many other areas, as shewn both by place-names, and by excavations, meres or lakes existed, it is apparent that much of the country would once be morass and impassable. It has been suggested that possibly the Romans or the Danes may have embanked or drained part of the area, but I am not aware of any direct evidence of this. It is, however, certain that the Britons took advantage of the marsh nature of the ground at Skipsea in arranging their protective earthworks at that place. Old maps and plans indicate the once marshy nature of our low-lying lands, and this was remarkably so in the vicinity of Kingston-upon-Hull, even so late as the seventeenth century, when the town could be isolated by allowing the water to enter the sluices to the east and west of it.

The question of drainage seems to have been very seriously and thoroughly considered during the reign of Edward I., towards the end of the thirteenth century and

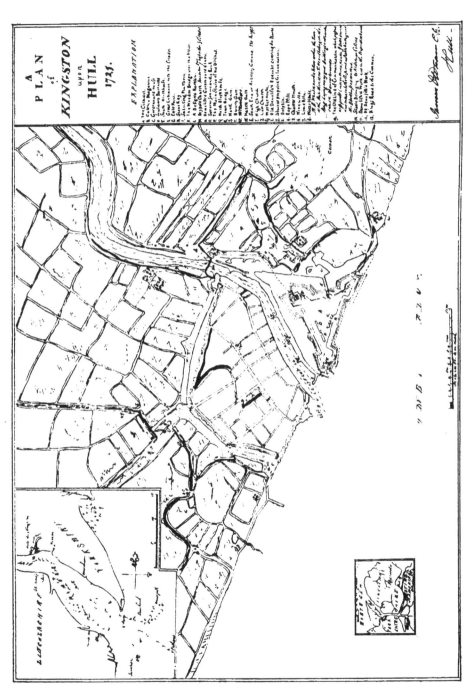

PLAN OF THE ENVIRONS OF HULL IN 1725, SHOWING THE NUMEROUS WATER COURSES ROUND THE TOWN.

early in the fourteenth. Many details respecting this are quoted by Poulson,* but the following extracts refer more particularly to the parishes adjoining the coast and the Humber. It is interesting to observe that many places are mentioned which have since entirely disappeared by the action of the sea ; while many others are referred to, the sites of which cannot be traced to-day, although they are upon *terra firma*.

We find that so early as 1285, Edward I. being informed that the land in Holderness was often " drowned " for want of repairing certain banks " on which the violence of the River Humber had made certain breaches," gave orders for the work to be attended to, and similar instructions were constantly repeated till 1387, the damage apparently principally occurring in consequence of faulty banks at Paull, Easington, Hedon, Frishmarsh, Thralesthorpe, Keyngham, Ryel (Ryehill), Burton Pidsea and Halsam, Patrington, Otrygham, Weynestede (Winestead), Frothingham (Frodingham), Newsum, Rymeswell (Rimswell), Outhorne, Wythorness (Withernsea), Redmayr (Redmere), Holaym (Hollym), Risum (Riseholme), Holmton, Thorpe juxta Wellewyk (Welwick), and Penysthorpe.

In a " Search " taken at Skipsea in the reign of Edward III. (1367), it was evident that certain drains or " sewers " required extending, that a highway through Routh was defective, and that Hull Bridge was also :—

" A certain sewer between Hornsea Burton and

* Vol. I., pp. 116-140.

Suthorpe . . . is insufficient, alsoe a certain sewer, called Water-dyke, in a field at Esington and lieth in length from Blyth Bridge into Frithomfleite . . . is defective. . . . Also, there is another sewer called Frithomfleite, and it lieth in length from Waterdike even to Freland Gote. . . . Alsoe, there is another sewer called Nookedike, and extends itself in length from Witholme even Coulaud. . . . Also, a certain high way called Stockbridge, which lies between Outhorne and Rimswell and is defective. . . . Alsoe, a certain sewer called Smalkedike and Outpitts, in the village of Holyme. . . . Alsoe there is another called Commerffleite which extends itself from the field of Winestead even to Potterfleite Haven . . . Alsoe, there is a certain wall of Humber on the village of Ottringham called Monasedike, and lies in length from Potterfleit Haven even to Mersknenladike, containing 47 cord, which is not repaired. . . . And a certain sewer called Bernardsleit lieth between Thorngumbaud and Thornemcroft, and the manor of Little Humber, lies in the Walls of Humber with one gote. . . . Alsoe, a certain sewer lies between Warkholme and Neuton Lands, beginning at Spelane Bridge, and goes even to Yowland Hirne, and is insufficient. . . . Alsoe, a certain sewer in the village of Ottringham, called Westcroft Dike, and is defective, beginning from South Land Hirne to the north part, even to Canufleit. . . . Alsoe, a certain sewer called Winestead fleit, extending in length from Byrstall Foss even to Humbersedike. . . . Alsoe, a certain way at Falait of Newton, near unto Yowcotes of Newton. . . . Alsoe, there is another high way called Burton Causey, and is defective. . . . Also there is a certain sewer beginning at Gildeson Marre, in the field at Tunstal, extending itself even to the demises of Hilston. . . . Alsoe another sewer beginning at Kingborough Crusting even upon the seashore and stretcheth itself between

Grimston and Kingborough even to the demises of Garton, and thence near Garton on the one part and Kingborough Newton and Aldborough on the other part even to the demises of Flytling. . . . And from the said demises of Flytling even to the demises of Flinton. . . . And from the said demise of Flinton, between Flinton and Flytling, even to Greenhilsikegote. . . . Alsoe, another sewer beginning at the demises of Humbleton, near the premises of Brakenhill, and extends itself between Flynton and Flytling, even to the aforesaid Greenhilsikegote, and is called Flytling Dike and Flynton Dike . . . .; and in the aforesaid Greenhilsikegote the aforesaid two sewers rune, whose courses stretch themselves even to Flynton Foss, and is insufficient. . . . Alsoe, there is a certain sewer beginning at the demises of Aldborough, and stretcheth itself even to the demises of Flynton, . . . And from the demises of Flynton even to the aforesaid Greenhilsikegote, where the aforesaid two sewers meet, . . . And from the aforesaid coarse there is two sewers, even to the demises of Humbleton, called Flynton Foss, and is insufficient . . . .; and from the said Flynton even to the east side of Water Millndam, called Humbleton, . . . And alsoe the sewers aforesaid viz., Flynton Foss and Humbleton Foss, from the aforesaid Greenhilsikegote to the aforesaid Water Milndam, . . . Alsoe, a certain sewer called Scryhildike, extending from the demises of Flytling to Humbleton Morskdike, . . . And from Mamberghmarr to the west bridge of Estranwick, even to Sandwath, and is insufficient, . . . And from the aforesaid Sandwath, even to Rugemunde Marr, is a certain sewer which is called Scurthdike, and is insufficient, . . . And from the aforesaid Rudgmund Marr even to Reedbuske, . . . And from the aforesaid Reedbuske, by the Bridge which is called Paytem Bridge, . . . And from West Carclute even to Parraknoke, . . .

And at Parraknoke the abovesaid water divides into two parts or courses, one of which extends itself from the aforesaid Parraknoke on the south part even to Rane-hokedike, and thence through Ranehokedike even to a close near Headon Fleit, . . . Then abutts another course which extends itself from the aforesaid Parrak-noke even to the field of Rihill, between Parrake of the Man'r of Burstwick and west carre of Bondsburstwick, . . . And it is to be known the aforesaid course of Parrakdike and Raneshokedike is very needful for the swift sliding of the water aforesaid, and the demises of the man'r of Burstwick being drowned, viz., Thaker, Doke, and Dramer, which other course is called Scurth-dike. And at the aforesaid close, near Headon, runs the two courses aforesaid, and one great sewer, called Headon Fleit, which extends itself through the middle of Headon even to Humber, and is insufficient, . . . Alsoe, there is a certain sewer through Waxam, Rymswell, and the middle of Smalsyn, and is defective, . . . Alsoe, another certain sewer called Keyingham Fleit, beginning at Crackhowe in the field of Tunstall, and extends itself even to Humber, . . . And it is to be known that the aforesaid of Keyingham Fleit is insufficient, and cannot be repaired unless they first repair that which ought to be repaired near Humber, Alsoe there is a certain sewer beginning on the east side or part of the village of Flynton, extending itself from the gate of Walter of Flynton, even to Moortofts, and thence even to the demises of Humble-ton, and is defective, . . . And the jury say that by the demises of the man'r of Humbleton ought to be repaired even to Twierdike, and is defective . . ."

Even so recently as 1772 an act was passed for draining a large area of 2500 acres or thereabouts between Ald-borough, Burstwick, and Ottringham, and to the south-east thereof, which " for many years past, have been frequently overflowed with water for want of proper drains and outfalls, and are thereby rendered in a great degree useless to the owners and a loss to the public ; and are also in danger to be further injured by the encroachments of the German Ocean."

Similarly, in 1796, at a meeting of the owners and proprietors of the low grounds and carrs in Keyingham Level, it was agreed that " in consequence of the great and increasing quantity of mud and warp in the channel, called the North Channel, from the places called No Man's Friend, Eastwood, and on the lands on each side of the said channel called Ottringham Growths, and Newsam or Sunk Sands," the drainage of the Keyingham Level had become ineffectual and should be extended.

Also, in 1802, by reason of the vast accretion of new land between Sunk Island and the shores, and the silting up of the channel, it was recommended that alterations be made in the draining by opening " a certain ancient drain called Hull Bridge Drain unto the River Humber, through or near to " Hedon Haven and Stone Creek.

CHAPTER VII

Spurn : its Formation—Reedbarowe's Light Tower
—Angell's Lights—Smeaton's Lights—Changes
in Spurn—Wilgils at Spurn—Leyland and Camden.

With one exception, the various capes and bays and
other irregularities on the Yorkshire coast can all be
explained on the theory of the survival of the fittest.
Flamborough Head, Filey Brig, and other points, are hard
masses of rock that have been left behind by the waves,
while the softer material on each side has been denuded.
The exception is Spurn Point. This, instead of being a
hard rocky promontory, as might be inferred from its
shape as shown on the map, is a low-lying bank of sand
and shingle, which for its stability largely depends upon
the roots of the marram grass, and groynes or other
artificial methods of protection.

Spurn Point owes its existence to the material carried
down the Yorkshire coast by the tides, which has fallen
on meeting the waters of the Humber. The southerly
and easterly directions of the waters of .the North Sea

and the Humber respectively, define its shape. Year by year the point extends southwards and westwards. The rate of growth can be ascertained by reference to plans, as well as (in recent times only) by actual measurement. On account of the shipping it is necessary that a lighthouse should be as near the extremity as possible.

Photo by]   [C. W. Mason.

NEAR SPURN POINT, SHOWING THE METHOD OF PROTECTION.

As the point grows the position of the lighthouse has to be altered.

The rate of growth of the point can be gathered from the following particulars :—

In 1428 Richard Reedbarowe, the hermit of the chapel at Ravensporne, obtained a grant to take toll from ships for the completion of a tower, which he had already begun as a lighthouse.

Photo by]   Spurn Point from the top of the Lighthouse.   [Godfrey Bingley.

In 1676 a patent was granted by Charles II. to a Mr.
Angell for the erection and maintenance of certain lights
at Spurn Point, which lights were erected at the request
of those interested in the northern trade, who represented

LETTER, DATED NOV. 5, 1675, FROM JUST. ANGELL, IN REFERENCE
TO THE SPURN LIGHTS.
(*The original in the Wilberforce Museum, Hull*).

that a broad long sand had been thrown up at the mouth
of the Humber a few months previously. Smeaton
thought that this sand had afterwards become connected
with the mainland, and so formed the Spurn Point of
his day. Greenvile Collin's chart of 1684 shews Angell's

Lights at Spurn. In the Wilberforce Museum at Hull is an interesting letter from Mr Angell, addressed to the Mayor of Hull. It is reproduced on the previous page.

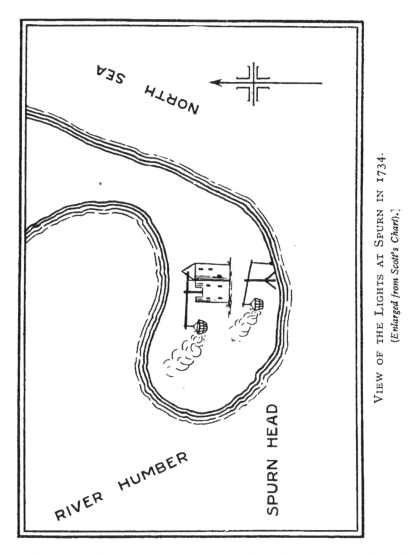

VIEW OF THE LIGHTS AT SPURN IN 1734.
(*Enlarged from Scott's Chart*).

After 1684 the point continued to increase in length, and the lights consequently became useless. An application was therefore made to Parliament in 1766, by the

Trinity House of Hull, for powers to erect and maintain other lights, which was passed. Smeaton was consulted, and he recommended the erection of two lighthouses.

In 1771 Smeaton reported that the point had extended 280 yards since 1766, and that it had increased on the sea-side to the extent of 50 yards. Many further facts

ELEVATION of the HIGH LIGHTHOUSE upon the SPURN POINT, & of the SWAPE by which the LOW LIGHT is Exhibited

THE OLD HIGH LIGHT AND LOW LIGHT AT SPURN IN 1786.
(After Smeaton).

relating to the subsequent changes will be given later, of which the following is a summary :—

Smeaton's small lighthouse was built in 1771, 280 yards east of the high light. A second was built 70 yards further west in 1816. A third was built 30 yards still further west in 1830. A fourth, 50 yards still further west, in 1831, and in 1863 the sea had reached the high light itself ; making a total westerly advance of 280 yards in 92 years, or 3 yards per annum. The low light in 1869 stood on the Humber side of the high light. The new lighthouse at Spurn is 71 yards north of that built in 1831.

According to Shelford, between 1676 and 1851, the southerly extension was 2530 yards, or 44 feet per annum. Between 1851 and 1888 the high-water line extended 600 feet in a southerly direction, or 17 feet per annum. During the same period, the westerly movement was 8 feet per annum on the North Sea side, and 17 feet per annum on the Humber side, a net yearly increase in the width of the point of 9 feet.

The foregoing information gives some idea of the rate of movement and accumulation of the sand and shingle at Spurn, but this represents only a very small proportion of the material washed from the Holderness cliffs ; all the clay and a great proportion of the sand and pebbles finds its way to the North Sea.

There is no doubt that a time will come when the flow of the waters of the Humber will prevent the further southern extension of the peninsula ; or, if the growth towards the Lincolnshire shore continues, a break must occur in the present sandbank. As it is, it is fairly evident that under favourable conditions pebbles, etc., are carried across the estuary to Lincolnshire. No doubt such changes as those referred to have frequently taken place. On more than one old chart an island is shown where the peninsula now is. Is it not possible that the old town of Ravenser, which existed on an island at the Humber mouth, but which was gradually washed away, may have had its foundations on a portion of Spurn Point at one of these critical periods of its career ?

So early as the seventh century we have a reference to Spurn, which Boyle found in Alcuin's *Vita Sancti Willibrordi*, printed in *Monumenta Alcuiniana* (Berlin, 1873).* Wilbrord, the great apostle to the Frisians, was born in Yorkshire, in 657 or 658. His father, whose name was Wilgils, late in life betook himself " to the promontories which are encircled by the ocean sea and Humber river," where he remained to the end of his days " in a little oratory dedicated to the name of St. Andrew, the apostle of Christ." His fame became noised abroad, and he was reported to have wrought miracles. Many resorted to his cell, and the king gave him " as a perpetual gift " certain small patches of land adjoining the promontory that a church might be built there. On his death his bones were laid in his " seaside cell."

Apparently one of the first historical notices of Spurn occurs in the report of an inquest in the third year of Edward I. (1273) in reference to the damage done to Grimsby by the forestalling practised by the men of Ravenser-Odd. " No doubt the ' oldest inhabitant ' came forward to give evidence, and it was through him that the hundreders were informed that some forty years ago, or more, the sea had at the entrance of the Humber cast up an island consisting of sand and stones, and that William de Fortibus, the then Earl of Albemarle,

* See "Trans. Hull Sci. and F.N. Club," 1899, p. 69.

had taken possession of it and commenced to build a town thereupon, the name of which was Ravenser-Odd. The date assigned to this event is circa 1233."*

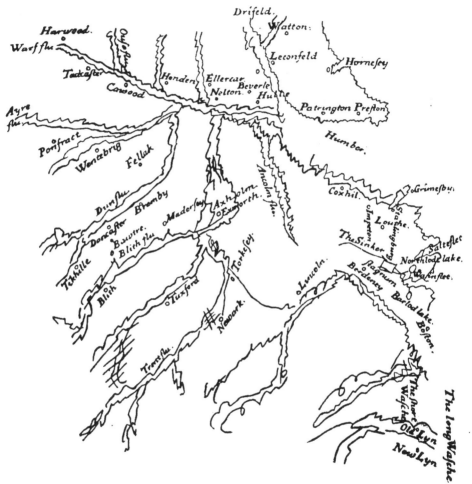

Odd was apparently an island in 1288. At the end of the fourteenth century it is referred to in the Meaux Chronicle as a peninsula.

* Lewis L. Kropf, " Hull and East Riding Portfolio," 1887, p. 71.

Leland, in his Itinerary, gives the earliest known map of the Humber, which, though crude, and obviously inaccurate in many ways, certainly shows no long sand-bank at Spurn.* Saxton's maps (1577) are apparently next in order of date ; though these are very rare, and his maps were copied by Speed a year or two later, and in that artist's " Theatre of Great Britain " (1611) they are fairly well-known.

In Camden's " Britannia '" (1586) it is stated that " on the very tip of this promontory [*i.e.*, Holderness], where it draws most to a point, and is called Spurn Head, stands the little village of Kellnsey."

* See " Hull and East Riding Portfolio," 1887, p. 107.

CHAPTER VIII

Smeaton's Narrative—Spurn Surveyed—Plans and Descriptions—Descriptions of Lights—Ancient References.

John Smeaton, in 1791, published an elaborate and valuable " Narrative of the Building . . . . of the Eddystone Lighthouse : to which is subjoined an appendix, giving some account of the lighthouse on the Spurn Point built upon a sand." This work contains some useful plates and much information bearing upon the changes at Spurn Point.*

In this he refers to the previously mentioned fact that in 1676 Charles II. granted to Justinian Angell of London a patent enabling him to continue, renew, and maintain certain lights at Spurn, at the requests of masters of ships interested in the northern trade. Mr. Angell, it appears, was the proprietor of the only piece of ground suitable for the purpose, and he erected two lights thereon. These are shown on the chart, see page 77.

Later, a " Broad Long Land "† became dry land at high water, and increased in extent so much, that prior to 1766, Angell's lights were by no means at the extremity

---

* The original drawings, etc., for this supplement recently came into my possession.

† Camden (1590) appears to suggest that in his time Spurn Point " seems to have been no more thar a sharp Head of land, that did not extend far from Kilnsey, and was thus called Spurn Head."

75

of Spurn, and consequently became useless. Parliament was therefore petitioned, and an Act was passed " as soon after the 1st of June, 1766, as conveniently could be " to erect two new lighthouses at or near Spurn Point. Smeaton was then consulted, and surveyed the district. He pointed out that one of the previously existing lights (the *low* light) was not a lighthouse in the ordinary sense of the word, but was upon a *swape* or lever, acted upon by hand, by means of which the light was drawn up in the

REVERSE OF MEDAL WITH VIEW OF SMEATON'S LIGHTHOUSE.

same way that water was in some districts drawn from deep wells.

In July 1766 Smeaton was instructed to make designs for two lights, 300 yards " asunder," the great lighthouse to be at the point, 90 yards from high-water mark, and 150 yards within the Spurn ; the low light to be 116 yards from high-water mark, without the Spurn.

In 1771, in surveying the Spurn, Smeaton found that the sea had so far encroached upon the east side that the very place fixed in 1766 for the low light was now on the high-water mark itself (see page 77). It was therefore arranged to build this light 80 yards more inland. Similarly, the high light was put 60 yards more to the north-west than was originally intended. It was clear that the Spurn sands were shifting westward,

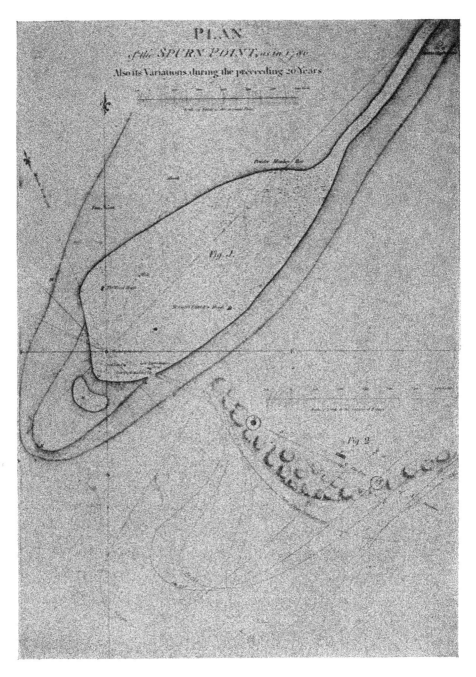

SMEATON'S MAP OF SPURN POINT IN 1786.

On Fig. 2 are shown the various positions assumed by the sandbank during the
preceding twenty years. The site of Angell's Lights is shown in the
top right-hand corner.

and after a storm in 1776 the coast was " taken away " so much that it not only destroyed the site of the old lighthouse, but laid bare the circular wall of the low light. On September 5th, 1776, the lights were kindled with *stone coal* or " cracklers " from the West Riding, which was better for lighting than the strong coal of Newcastle and Sunderland, and this stone coal " exhibited an amazing light, to the entire satisfaction of all beholders."   The fires were protected by lanterns, instead of being bare, as on the previous lights.

In 1780, Smeaton surveyed Spurn Point again, and found that the old lighthouse foundation, which in 1771 was wholly within the unbroken land, was now 50 yards without the present border, toward low-water mark. Similarly, the land where the lighthouse stood in 1771 had a considerable breadth, but " now it lies opposite to the narrow *isthmus*, which, after extending about a quarter of a mile to the north-eastward of this site, becomes a naked beach or ridge, over which the sea breaks into the Humber " and it extended about half a mile in that form.

There are one or two interesting plates accompanying Smeaton's report. One is a map, of which the following description, written well over a century ago by a practical engineer, is not without interest :—

" It appears that this coast, from *Flamborough Head,* or at least from *Bridlington* to *Spurn Point,* trending south-south-east . . . and the tide of the flood of the German Ocean setting strongly to the southwards ;

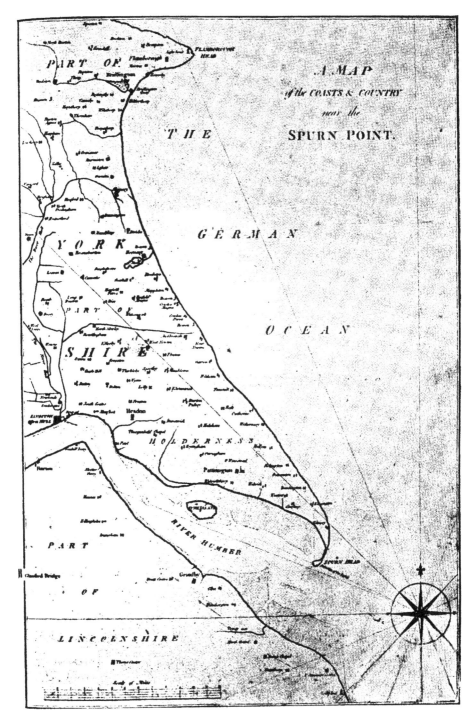

SMEATON'S MAP OF EAST YORKSHIRE, 1780

this will cause these flood-tides, when agitated by all winds from north-north-west to north-east, to *bite* very hard upon the *stretch* of coast, and, no part of it being rocky, to *wear away*; the sand and matter dislodged to be driven towards the south, forming at the tail of the land, the *appendage*, called the Spurn Point. Doubtless this matter so brought would in time *block up* the *Humber*, was it not for the *powerful re-flow* of that river's tide, aided by the fresh water from the higher country. The *Spurn Point* being therefore the effect of a *struggle* between the *sea tide* of the *German Ocean* and the *re-flow* of the *Humber*, we are not to wonder if the powerful effort of the sea by degrees drives the *channel* of the *Humber* southwards towards the *Lincolnshire coast*, thereby giving opportunity for the Spurn Point to *lengthen* towards the south, and, as the coast wears away, to which it *hangs* as a *rudder*, to be also in a state of *travel westward*."

This map, which measures 12 × 19 inches, and is on the scale of 3 inches = 8 miles, was engraved by W. Fayden, Geographer to the King, 1780. It shews beacons at Hornsea, Cowden, Awdbrough [Aldbrough]; one between Rudston and Burton Agnes; and decoys are marked at points not defined by place-names, but probably represent those at Scorborough and Meaux.

The Plan of the Spurn Point, as in 1786 (see page 77), also its variations, during the preceding twenty years, is described as under, by Smeaton :—

" Fig. I. is the *Plan* of the *Spurn Point*, as taken in the year 1786, comprehending about 98 acres; upon which little needs be said, after what is upon the *face* of it.

Fig. II. is an enlargement of the *extreme* point of the *Spurn*, wherein the boundary line of the *Sand Hommacks*, and its relative *high water line* is marked with the *year*.

The space comprehended between these two lines was a kind of flat area or *fore shore* over which the sea beat in *rough weather*; and upon which (the whole being then considered as *rapidly increasing*), the *lower light* was proposed to be built at *A*; ANGEL'S *lighthouse* being at that time in use; and within the boundary of the *firm part* of the *peninsula*. *B* was the place marked out for the *high light*; and parallel to *A B*, the proposed *line of direction*, at the distance of 23 yards to south-westward, were the two points *C, D* marked out for the *temporary lights*; but for reasons that appeared before their erection, the point *C* was in reality carried to *E*. *F*, a low building, containing *two cottages* for the temporary *light-keepers* was erected at the same time. When the lighthouses were begun in 1771, the high-water line was as marked for that year. For *reasons* the *low light* marked at *A* was placed at *G*. The *high light* house was begun the latter end of the year 1772, and placed at *H*. In 1773 the high-water line was as marked, and the boundary of the *Sand Hommacks* remained nearly the same for some time; but a *great storm* in *January* 1776 varied the high-water line, as marked for that year. After that there was little variation of that line to the *conclusion* of the work in 1777.

" On visiting the *Spurn* in 1786, the *high-water line* ran as described for that year, approaching near to the *low light machine* at *Q—I.K.L.* Fig. I. is the high-water line, and *M.N.O.* that of low water; from which it will appear, that the *Spurn Point* marked *P*, is advanced considerably to south-westward from the line of direction, and has considerably retreated to north-westward, or nearer *a-breast* of the *high-light*. Of the *island R*, there were no traces in 1777; but now, being grown with the *bent grass* the *sand hommacks* had established themselves, and rendered the surface irregular. It appears that *little change* has happened near the *high lighthouse*, but that the *breadth of sand* was there increased."

The plan is 13½ by 19 inches, and was " Engraved by W. Faden, Geographer to the King, 1786."

Fig. I. is at the scale of 133 yards to an inch, and well shows the shape of the southern extremity of Spurn a century and a quarter ago. It will be seen that the site of Angell's lighthouse is in the north-east corner.

Fig. II. is twice the scale of Fig. I., viz., 266 yards to an inch. It shows the sand-dunes and the positions of the lights in 1766, and the various dotted lines indicate the remarkable changed positions of the Point in the years 1771, 1773, 1774, 1776, and 1786 (see page 77).

The other plate (page 70) shows the high lighthouse, built by Smeaton, with its surrounding wall, which latter still remains at Spurn ; and the " Swape " by which the low light was exhibited. It will be seen that the swape was built on the see-saw principle ; the iron fire-basket for the reception of the coal being lowered to be filled and lighted, and then lifted up into its position. This engraving is at the scale of 1 inch to 12 feet, measures 12 by 18 inches, and is by J. Record, 1785.

There is evidence of the continued change of the position of Spurn. Shelford* says that between 1676 and 1851 the movement was 2,530 yards south, or 14½ yards a year. Mr. Butterfield† finds that from 1851 to 1888 the high-water line has advanced 200 yards in a southerly direction, or nearly 6 yards a year. Thus

* Outfall of the River Humber. " Proc. Inst. C. Eng.", Vol. XXVIII., 1869.
† " The Naturalist," November 1904, p, 326.

between 1676 and 1888 there is a recorded growth of land at Spurn of 2,730 yards.

Also, in addition to moving in a southerly direction, it has been growing in width, and extending westwards. Between 1851 and 1888, the westerly movement on the North Sea side has been 99 yards, or 8 feet a year, whereas in the same 37 years, the westerly movement on the

THE SPURN LIGHTS IN 1829.
(*From Allen's " Yorkshire "*)*.

Humber side has been 210 yards, or 17 feet a year; a gain in the width of the land of 111 yards.*

According to Shelford and Pickwell, the average rate of the known southward extension of Spurn Point for the 200 years prior to 1875 has been 13½ yards per annum.

In view of the extraordinary changes in the position

* Butterfield, " Naturalist," 1904, p. 328.

of the Spurn as shown by Smeaton and others, the question
of the original position of Ravenser may not be so simple
as at first considered.   It is of course usually assumed that
it was to the west of the present Spurn.   The late Thos.
Blashill suggested, however,* that the site of the lost
town of Ravenser was outside, *i.e.*, to the east of the
present Spurn Point.†

At that time there seemed little direct evidence that
an island ever existed to the east of the Spurn.   Since

PRESENT LIGHTHOUSE AT SPURN AND THE OLD LIGHTHOUSE,
NOW IN THE WATER, AND USED AS A MAGAZINE.

then, however, while examining various MS. maps in
the National collection, with the kind help of the officials
at the British Museum, I was able to see a parchment
of the time of Henry VIII.   This shewed the Humber,
with its sandbanks, etc., but the principal point of interest
is, that an island is distinctly shown to the east of the then
Spurn Point.   This plan does not appear to have been
seen by previous writers, and by the kindness of the

* " The Naturalist," September 1904, p. 264.
† See also the map accompanying Shelford's paper.

84

Photo by] THE OLD LIGHTHOUSE, SPURN. [R. Fortune.

British Museum authorities I am able to give a reproduction of it (see page 217).

In 1622, Collis, in his " Lectures on Sewers " recorded
that " of late years parcel of the Spurnhead in Yorkshire,
which before did adhere to this continent, was torn
therefrom by the sea, and is now in the nature of an
island."

As already pointed out, this sort of thing doubtless

THE HOUSES AT SPURN, SHOWING METHOD OF PROTECTING
AGAINST EROSION.

frequently happened, as it has in more recent years. So
recently as on Hewitt's chart of 1828, a portion of the
Spurn neck, north of the lighthouses, is marked as
" overflowed at high spring tides." In 1849 a serious
breach occurred, and groynes were then erected to protect the sandbank.

CHAPTER IX

Lost Towns : Orwithfleet, Tharlesthorpe, Burstall, Ravenspurn—Alleged View of Ravensere—Ships for the King's Navy—Ravenser-
Odd : its Ruin—References in the Meaux
Chronicle.

An early record of the loss of land in south-east Yorkshire occurs in the Meaux Chronicle. In this, reference
is made to a suit instituted by Sir John of Meaux against
the convent for refusal to pay rent on 33 acres of grass-
land in Orwythfleet, which were carried away by the
waters of the Humber between 1310 and 1339. Orwithfleet was apparently on the Humber shore, to the west of
Easington.

Tharlesthorp yielded to the monks of Meaux in 1246
300 quarters of grain, principally corn. In 1277 there
were 1274 sheep at pasture in Tharlesthorpe, and the land
is represented as being so rich that the ewes generally
brought two lambs*. The monks of Meaux also had right
of pasturage on " The Green " there. In 1336-7 Sir
Robert Constable of Halsham, died, " seized of one wind-
mill, eight tofts and four bovates of land in Tharlesthorp."†
In 1342-3 Ralph de Bulmere and others were appointed

* See " Trans. East Rid. Antiq. Soc.", Vol. I., 1893, p. 27.
† " Poulson," Vol. II., p. 528.

to repair the banks at Tharlesthorpe, Frismersk and other places on the Humber side.

In the fourteenth century the land at Tharlesthorpe, in common with that at. Frysmerk, Saltagh, Wythfleet, and Dymelton [Dimlington] and Ravenserodd, suffered so much by the sea that its value fell from £250 a year to £50 a year.

Soon after, " the sea made further encroachments,

BUCK'S VIEW CF BURSTALL PRIORY, AS IT WAS IN 1721.

Buck gives the following description of " Burstall Abbey ":—" Stephen, Earl of Albamarle, nephew to Wm. Rufus, gave a Cel, with Lands and Revenues, to the Monks of St. Martin, without the Castle of Albamarle, in Normandy, and Walter, Archbishop of York, setled these Benedcts at Birstal with Immunities, A.D. 1115. Lastly the Abbot and Convent, granted by their Deed (dated 18th Richd. 2d, 1394) all their Lands, Tithes & Pensions, in England, to the Abbot and Convent of Kirkstale in the same County."

so that only a third of the original estates of Tharlesthorpe (about 90 acres) could be saved, and that by the erection of a costly sea-wall." This was during the reign of Edward III.* In 1353-6 it is apparent the work of destruction by the waters of the Humber commenced,

* Boyle : 1889, p. 71.

88

and concluded by the place disappearing altogether about half a century later.

Burstall, or Birstal, is another place that has suffered considerably through the action of the waters of the Humber. "Burstall Priory" was figured by Buck in 1721, but it was then a ruin and apparently had been for some time. Poulson gives a wood-cut * as an initial letter to his chapter on Skeffling, in which parish Burstall

BURSTALL PRIORY.
(*After Poulson*).

is situated. The remains of the building, which was an alien priory, have now entirely disappeared; though many of the stones are apparently to be found in the works erected to protect the shore of the Humber at this point. "The first ordination of this house or cell at Byrstall, is dated at Beverley, in the month of June, 1219." Several Saxon, etc., coins and other objects have been

* Evidently copied from Buck's view.

found on the site of this place, and Poulson (Vol. II., p. 505) figures a massive Roman bronze brooch which was found there. This writer states that "the priory of Burstall is swept away by the frightful encroachments of the sea ; and from the numerous relics and fragments of other times washed upon the shore below Welwick, it is conjectured that this must have been the site of a populous place."

Burstall Priory is shown on the Humber side, almost due south of Skeffling, on Tuke's map of 1786, (see p. 233).

In the adjoining parish of Welwick were Penisthorp and Orwithfleet, which have now disappeared by the inundations of the Humber.

William Shelford's paper* is one of considerable value in connection with our enquiry, notwithstanding the fact that, with the exception of Smeaton, he considers that "all historians of the locality have evidently been conscious of their inability to deal with the physical causes of the events which they are recording."! He points out that Spurn Point, even in Roman times, must have been 2,250 yards at least beyond the present coastline ; and that at or near this spot the Danes landed in 867, planted their standard "The Raven," and practically originated the town of Ravensburg, or Ravenser, or Ravenseret, within Spurn Head. The town developed into "one of the most wealthy and flourishing ports of the kingdom. It returned two Members to Parliament, assisted in equipping the navy, had an annual fair of

* Proc. Inst. C. Eng., 1869.

thirty days, two markets a week, is mentioned twice by Shakespeare,* and considered itself honoured by the embarkation of Baliol with his army for the invasion of Scotland in 1332 ; by the landing of Bolingbroke, afterwards Henry IV., in 1399 ; and by the landing of Edward IV. in 1471, not long after which it was entirely swept away."

To-day, we cannot even be certain where the place was. The cross now at Hedon, is said to be from there, and also is said to have been erected as a memorial to the historic landing referred to, by one " Martine de la Mare." Two church bells, one at Easington, and one at Aldborough near Hornsea, are likewise said to be from Ravenspurn. These are all apparently that exist from this place. We know a Chapel of Ease was built, and was in existence as early as 1272 ; and we know that it had a street called Locksmith Lane.†

An alleged " fifteenth century illuminated manuscript " contains the only view of Ravensere in existence. The view represents a church at the end of a fairly wide street, with a row of houses on each side. Towards the centre of the street is a cross. It appeared in an anonymous article on " The Lost Land of England " in the " Strand Magazine " for October, 1901. On communicating with the publishers, Messrs. Newnes, they inform me they are not able to say where the illustration was taken from. The article, however, was evidently

---

* King Henry VI., part iii., Act iv., Scene 7; and Richard II., Act 2, Scene 1. † " Ocellum Promontorium," p. 270.

written by the same author who wrote " The Story of Lost England," * viz., Mr. Beckles Willson. Mr. Willson was written to, but regretted he could not remember the source of his illustration. This is particularly unfortunate, as no other view of any part of this lost town appears to be extant. From the unusual width of the street shown on the " illuminated

ALLEGED VIEW OF RAVENSPURNE.

(Probably a forgery).

manuscript "; from the positions of the houses and the church, and also the cross—supposed to be the first view of that now at Hedon—I have my doubts as to

* Newnes, 1902. In this later work the view of Ravensere is not reproduced nor is it referred to in any way. The artist has evidently taken his idea from Poulson's view of Sutton church, which is reproduced on page 139. A comparison with this, and the alleged view of Ravenspurne, shows the extraordinary similiarity of this alleged fifteenth century MS. and Poulson's drawing of 1841. It will be observed that the initial letter " I " in the one is transformed into a market cross in the other l

the authenticity of the sketch. On enquiry at the British Museum, I find there was no such manuscript there, and the authorities agree with me that the illustration " never came from any 15th century illuminated manuscript in the British Museum or elsewhere." It would appear, therefore, that we have yet to see a view of any part of Ravenspurne.

Amongst other items of interest in connection with Ravenser, we find that in 1296 " Kaiage " * was granted to the inhabitants by Edward I.† Two years later Ravenser petitioned the king for certain privileges, and offered 300 marks in payment. In 1300 the magistrates of Ravensere were enjoined to stop the export of bullion ; in 1305 it sent Members to Parliament. In 1310 Ravensere remonstrated against the depredations of the Earl of Holland, and in the same year Ravenness sent ships for Edward II.'s expedition to Scotland. Two years later the inhabitants were empowered to levy a tax to defend their walls. In 1323 commissions were issued for the " Wapentak of Ravensere." In 1335-6 warships of Ravensere are referred to, and in 1341 Ravensere sent one Member to " a sort of " naval Parliament of Edward III. In 1346 one ship only was sent by Ravenser to the siege of Calais ; (Hull sent 16). In 1355 bodies were washed out of their graves in the chapelyard at Ravenser. In 1361 the floods drove the merchants to Hull and Grimsby ;

* Kaiage is evidently the old form of *Key*age or, to be more correct *Quay*age, evidently somewhat similar to the present "wharfage" charges at our sea-port towns.
† Phillips: Rivers, etc., of Yorkshire, p. 280.

and by 1390 nearly all trace of the town, as such, was gone.* In 1413 a grant was made for the erection of a hermitage at *Ravenscrosbourne,* and in 1428 Richard Reedbarowe, the hermit of the chapel of Ravensersporne obtained a grant to take tolls of ships for the completion of his light-tower. In 1538 Leland refers to Ravenspur in his " Itinerary," which seems to be the last reference to the place. As pointed out elsewhere, the place is not included in Holinshed's " List of Ports and Creeks," which was issued before 1580.

It is interesting to note in this connection that on the map accompanying Mr. Shelford's paper of 1869, it is clearly indicated how the position of Ravenser may well have been on the east of the present Spurn Head, though originally on the west of the Spurn of its time. Shelford gives the probable position of the Holderness coast-line in the ninth century, assuming that the rate of erosion has been fairly regular, and at $2\frac{1}{4}$ yards a year, which is a very reasonable estimate.

Ravenser-odd (also referred to as Odd near Ravenser, Ravenserot, Ravensrood, Ravensrodd, Ravensrode, etc.), probably originated in the early part of the thirteenth century, soon after Ravenser, the adjoining port, came to be of importance. Ravenser-odd was apparently built on an island.

In 1251 some monks obtained half an acre of ground on which to erect buildings for the preservation of fish, in the burg of Od near Ravenser. The chronicler

* Frost's " Notices."

of Meaux wrote that " Od was in the parish of Esington, about a mile distant from the mainland. The access to it was from Ravenser by a sandy road covered with round yellow stones, scarcely elevated above the sea. By the flowing of the ocean it was little affected on the east, and on the west it resisted in a wonderful manner the flux of the Humber."

In 1273 there was a dispute about a chapel at Od, and this was carried on for some time.

In 1300 Edward I. gave some lands in Ravenserodde to the convent of Thornton in Lincolnshire, and others to St. Leonard's Hospital, York.

In 1315 the burgesses of Ravenserod agreed to pay the king £50 for the confirmation of their charters, and " Kaiage " for seven years. In 1326 the king granted dues and customs in the port of Ravenserod, and about 1336 William De-la-Pole left Ravenserod for Hull. Ravenserode sent a representative to Edward III.'s " naval Parliament " in 1344, as well as a man well versed in naval affairs.

In 1346 Ravensrodde was one of the places mentioned by the Abbot of Meaux as suffering by the sea. In the following year it was frequently inundated, and in 1360 " Ravenser Odd was totally annihilated by the floods of the Humber and inundations of the great sea."

In 1355 the bodies in the chapel yard, which, " by reason of inundations were then washed up and uncovered," were removed and buried in the churchyard at Easington.

About this time we read the following curious note in the Meaux Chronicle :—" When the inundations of the sea and of the Humber had destroyed to the foundations the chapel of Ravenserre Odd, built in honour of the Blessed Virgin Mary, so that the corpses and bones of the dead there buried horribly appeared, and the same inundations daily threatened the destruction of the said town, sacrilegious persons carried off and alienated certain ornaments of the said chapel, without our due consent, and disposed of them for their own pleasure ; except a few ornaments, images, books, and a bell which we sold to the mother church of Esyngton, and two smaller bells to the church of Aldeburghe. But that town of Ravenserre Odd, in the parish of the said church of Esyngton, was an exceedingly famous borough, devoted to merchandise, as well as many fisheries, most abundantly furnished with ships and burgesses amongst the boroughs of that sea-coast. But yet, with all inferior places, and chiefly by wrong-doing on the sea, by its wicked works and piracies, it provoketh the wrath of God against itself beyond measure. Wherefore, within the few following years, the said town, by those inundations of the sea and of the Humber, was destroyed to the foundations, so that nothing of value was left."

Notwithstanding this, " In the Hedon inquisition of January 1401, the chapel of Ravenserodde, with the town itself, was declared to be worth, in spiritualities, more than £30 per annum."*

* Boyle : "Lost Towns of the Humber," p. 49.

CHAPTER X

KILNSEA—ITS LOST CHURCH—LAND LOST—MEASURE-
MENTS—RELICS OF THE OLD CHURCH—KILNSEA
CROSS—SUNTHORPE.

KILNSEA, the Chilnesse of Domesday, is feeling the
effects of the sea more forcibly than is any other place on

KILNSEA CHURCH IN 1829.
(From Allen's " Yorkshire ")

the coast. Reid records that in 1822 there was a
church and thirty houses ; on the ordnance map of
1852 there are still six or seven houses shown, and the
foundations of the church at half-tide. Since then
every brick has gone.

97   G

Between 1766 and 1833 Pickwell estimates the loss at 1.8 yards a year ; between 1833 and 1876, 5 yards a year ; and between 1876 and 1881, 3.3 yards, a total of 350 yards in 115 years.

STONE BUILT INTO THE WALL OF THE BLUE BELL INN, KILNSEA.
IT WAS 534 YARDS FROM THE SEA IN 1847.

On the Humber side of Kilnsea also the parish is suffering, though not to so serious an extent.

Many interesting facts are given by Mr. J.

Backhouse in a paper on "A Vanishing Yorkshire Village.*

The "Blue Bell" Inn at Kilnsea was built in 1847, and, according to the record inserted into the wall, was then 534 yards from the cliff. In 1852 it was 527 yards away; in 1876 it was 392 yards away; in 1888, 377 yards. In 1898 Canon Maddock measured, and found it to be

Photo by]   [Oxley Grabham.
"HOLY WATER STOUP" SAID TO BE FROM OLD KILNSEA
CHURCH.

328½† yards away; and Mr. Backhouse (1908) measured it as 200 yards; though an estimate in 1910 ("Naturalist," 1910, p. 342) gives the distance as 272 yards. This, however, was along the road which may not have been the nearest point to the cliff. Thus 334 yards of land

* Rep. York Phil. Soc. for 1908, pp. 49-59.
† Butterfield, "Naturalist," 1904, p. 326, says 333 yards in 1898, but he may have measured early in the year.

have been washed away at this point in sixty years, or 5½ yards a year. At Kilnsea Beacon, 177 yards were washed away between 1852 and 1886, and a further 50 yards between 1856 and 1898, a total of 227 yards in 46 years.

Some measurements taken in 1833 are given by Poulson (II., p. 522), though as most of the places mentioned have since disappeared, we are not able to compare

KILNSEA CHURCH AT THE CLIFF EDGE IN 1826.
(*After Poulson*).

them with present-day measurements. But it is perhaps as well to place them on record :—" In 1833 the south end of Kilnsea sea bank top 48 yards ; large farmhouse front door to the sea, 58 yards ; middle of the street in Kilnsea to the cliff, 47 yards ; churchyard gate to the edge of the cliff, 25¼ yards ; remains of the west end of the church only 4 yards ; distance of the same west-end to the widest extremity of the churchyard, 25½ yards ; middle of the road entering Kilnsea, below the hill, 10 yards. In 1766 the chancel of Kilnsea church was distant from the cliff, 95 yards ; suppose the church 30 yards in length, and 4 yards are yet remaining (1833), there is lost 121 yards ; but the large masses of stonework have preserved the foundations at least 4 yards, so that the waste is 125 yards, or more, in 67 years, on each side the ruin. These measurements are from the Rev. Jos. Hatfield, curate of Sproatley. It is estimated by Mr. Little, that from 1767 to 21st May, 1828, when part of the steeple fell, from a memorandum of Mr. Hunter, many years a resident farmer, has annually lost nearly 3 yards."

The fine stone cross said to have been erected by one " Martini de la Mare" at Ravenspurne, and supposed to be in memory of the landing there of Edward IV., or possibly of Henry IV. " was afterwards removed to Kilnsea, where it stood for many years until removed to Burton Constable, and finally was again removed to Hedon, where it now exists."

Mr. Backhouse records that Philip Loten, father of the Easington naturalist of that name, was born in old Kilnsea, and remembered a road on the seaward side of the old church. As a youth he frequently played on the ruins of that church. The last person buried in old Kilnsea churchyard was a shipwrecked negro, in 1823 ; after which year services in the church were discontinued.

There are many relics of the old Kilnsea church, a list of which Mr. Backhouse gives in his paper. Among

THE OLD KILNSEA CROSS, FROM CHILD'S ORIGINAL DRAWING, 1818.

(By the courtesy of the Yorkshire Archæological Society).

THE OLD KILNSEA CROSS, FROM CHILD'S ORIGINAL DRAWING, 1818.

(By the courtesy of the Yorkshire Archæological Society).

them is a portion of a dated memorial tablet (1728), which he gave to the Hull Museum.

According to Phillips,* in 1833 the gate leading out

THE OLD KILNSEA CROSS AS RE-ERECTED AT HEDON.

of Kilnsea north field was 327½ yards from the cliff; the top of the south end of Kilnsea sea bank was 48 yards;

* " Rivers, Mountains, and Sea-coast of Yorkshire," p. 283,

the front wall of the large farmhouse, licensed for divine service, was 58 yards ; the old churchyard gate to the edge of the cliff was 25¼ yards ; there remained 4 yards of " ruin of the steeple," and from the said ruin to the extremity of the churchyard was about 95 yards. All this less than eighty years ago, and all gone. The last part of the tower fell in 1831 ; the first half having fallen five years before. Probably the last view of any portion of the site of old Kilnsea Church was when the East Riding Antiquarian Society visited the district in 1899, and were able to get well out at the lowest part at a very low tide. The remains were at a distance of 250 yards from the then cliff top.* In 1776 the church at Kilnsea had 95 yards between it and the sea.

The old cross, now at Hedon, was 50 yards from the cliff in 1790 ; whereas in 1833 its site was 30 yards seaward from the cliff ; thus 80 yards had disappeared in 43 years.

Two fields on the cliff at Kilnsea measured, in 1760, 30½ acres and 9¼ acres respectively. In 1827, they measured 23¼ and 6 acres respectively. In these two fields alone, 10½ acres were lost in 67 years.† Records from Tennison's farm, at the north of the village, between 1840 and 1876, show that the loss there has been 5.3 yards a year.

The map reproduced on page 107 shows more graphically than words how much of Kilnsea and its

* E. R. Ant. Soc., vol. 7, 1899, p. xxiv
† Phillips : " Rivers, etc., of Yorkshire," p. 284

neighbourhood has " gone to the sea " during the past century.

In early times there was evidently a creek at Kilnsea, of sufficient size to accommodate at any rate smaller craft, as is shewn in some of the early engraved maps. Holinshed, about the middle of the sixteenth century, issued a " list of such ports and creeks, as our sea-faring men do note for their benefit upon the coasts of England," and includes " Kelseie Cliffe," as well as Horneseie becke, Sister kirke [Withernsea], Pattenton-Holmes [Patrington Holmes], Kenningham [Keyingham], Pall [Paull], Hidon, Beuerlie [Beverley], Hull, Hull-brige, Hasell [Hessle], etc.

On Lord Burleigh's Chart (Temp. Henry VIII., see p. 208), the following note appears, opposite "Kilnsea" :— " In calme wether ships of good Burden may ride and lande here to do annoyaunce to the contreye."

As to the possibility of high cliffs formerly existing at Kilnsea, Thompson, in an " Appendix to his History of Swine " (1824, p. 233), says " Smeaton, the engineer, noticed the " *high clay cliffs* " of Kilnsea about fifty years ago [*i.e.*, c. 1775], when he visited Spurn Point ; and there is no doubt that the land eastward from Kilnsea was formerly of considerable height above the town. A clergyman who had lived to old age in that part of Holderness, and died not many years ago, was often heard to assert that he remembered a field, called *east-field*, lying between Kilnsea and the sea, which greatly rose in height towards the sea, but no east-field can now [1824] be found, and there is no

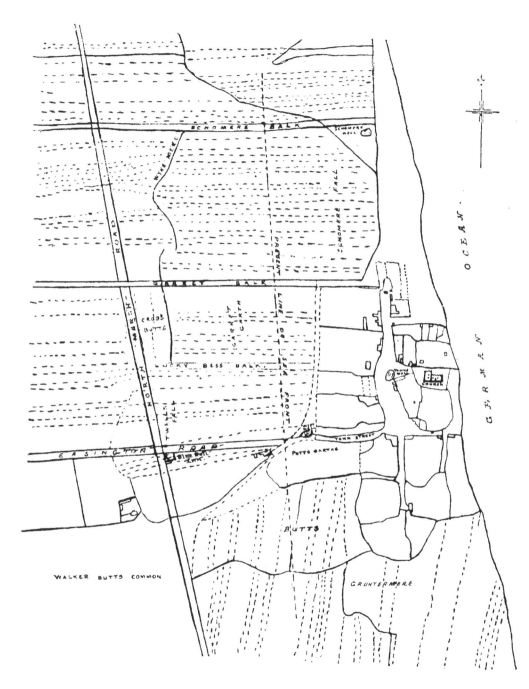

PLAN OF OLD KILNSEA, SHOWING AREA WASHED AWAY.

(From " A History of Wuthernsea ').

doubt that it was all swept away by the sea before the end of the last [eighteenth] century." Poulson, in the extract already given, refers to the road in Kilnsea village " *below the hill.*"

Thompson considers that this high land was really the original Spurn *Head*—the later and present Spurn *Point* being the sandheap to the south. In support of this theory he suggests that the Domesday name *Chilnesse* referred to the Chil nesse, or nose, or promontory, and that as the high land was washed away, and the *ness* disappeared, the name became altered to Kiln*sea*.

To further prove this he quotes Drayton's poem, *Poly-Olbion* [written 1612-1622].—

> " From Kilnsey's pyle-like point, along the eastern shore,
> And laugh at Neptune's rage, when loudest he doth roar
> Till Flamborough jut forth into the German Sea "

from which it seems clear to Thompson that " by *Kilnsey's pyle-like point* is undoubtedly meant, the high point of Kilnsea cliff, or Spurn *Head*." Kropf opines that Drayton merely had Camden's " Britannia " for his information, and had not actually visited this coast, and is therefore not reliable.

Smeaton in 1771 estimated that the loss of land at Kilnsea was 10 yards a year.*

" Practically the whole of the village of Kilnsea has been swallowed up during the past century. The encroachment has been at the rate of 4 yards per annum for many years, with the exception of the past twelve

* See his " Narrative of Eddystone Lighthouse," p. 188.

years, during which the average encroachment has been 3 yards per annum."* Groynes have now been erected here by the Board of Trade, with satisfactory results.

In 1650 Kilnsea Church was considered to be " well scittuate . fit to be continewed " (in the Parliamentary Survey of the Benefices of the East Riding).†

An early reference to " Kellnsey " occurs in Camden's " Britannia," where the village is described as " on the very tip of the promontory " of Holderness.

In Allen's‡ time (1831) it was expected that the ruins of the church tower would " probably exist for a considerable length of time ; the fallen ruins having made a strong bulwark against which the violence of the waves can vent itself without injuring the shattered tower." The final crash actually occurred in the same year that these words were printed!

We learn from a very scarce and scurrulous work, " The Churches of Holderness," by " Geoffrey de Sawtry, Abbot " (1837, p. 11) that " The church has long since been swept away ; and the tower, which stood many years after, a valuable landmark for seamen, fell with a tremendous crash, in the autumn of 1830. This is therefore another churchless village ; but having a population of nearly two hundred, they have set apart a room for divine service, in which it is performed every third Sunday, weather permitting ;

* Matthews, in 1905.
† " Trans. East Riding Antiq. Soc.," vol. 4, 1896, p. 56.
‡ " History of Yorkshire."

otherwise, it is reported, the worthy pastor, feeling for his flock, grants them an indulgence to remain indoors,

NEAR KILNSEA BEACON, LOOKING SOUTH: SHOWING THE FLOODED AREA, 1906.
(*From a Painting by Gordon Home*).

and takes the same himself." For many years, the bell, which was dated 1700, was suspended over a beam in a stackyard, and was struck by throwing stones at it!

In 1905-6 the low lands from Kilnsea Warren nearly

FLOODS AT KILNSEA: THE BEACON IN THE DISTANCE.

to Easington Lane end on the north, and beyond Skeffling on the west, were flooded in consequence of the sea breaking through the artificial banks along the coast. The sea-water remained on the land for a long time, the crops were all destroyed, wells were filled with salt-water, roads became impassable, and in many ways the inhabitants of this part of the country had very troublous times. From the appearance of the district it almost seemed that "history had repeated itself," and that in this neighbourhood the villages were once more to be

THE ROAD UNDER WATER AT EASINGTON, 1906.

" blotted out and consumed." The area inundated is shown on the map which appears as a frontispiece.

In a copy of the Cartulary or Book of Meaux, the village of Sunthorpe is said to be in the parish of Kilnsea, and, according to Thompson,* " it no doubt stood between Kilnsea and the Sea."

Boyle places it west of Kilnsea (in his " Lost Towns of the Humber ").

* Appendix to " History of Swine," 1824, p. 234.

CHAPTER XI

EASINGTON—DOMESDAY AREA—PRESENT AREA—LOSS
BY SEA—FLOODS—NORTHORPE, HOTON AND TUR-
MARR.

ACCORDING to the Domesday Survey " In Esinstone,
Morcar had fifteen carucates of land to be taxed ; and

THE FLOODS AT EASINGTON, 1906.

there may be there as many ploughs. . . . Drogo
has now there one plough, and thirteen villanes, and four
bordars ; having three ploughs and one hundred acres
of meadow. To this manor belongs the soke of these—
Garton and Ringheborg [Ringborough] eight carucates
of land to be taxed ; and there may be as many ploughs
there. Baldwin has now of Drogo himself there one

112

THE FLOODS AT EASINGTON IN 1906

THE FLOODS AT EASINGTON IN 1906.

H

plough. There is a priest and a church there, and sixty acres of meadow."

According to Boyle, the acreage in Domesday was 2400. In 1880 there were 1300 acres. in Easington, a loss of 1100 acres between 1086 and 1800.

From a passage in the Meaux Chronicle it is apparent

Photo by]   *[Oxley Grabham.*

EASINGTON CHURCH AND OLD TITHE BARN.

that Easington possessed a haven in the fourteenth century. It is there recorded that Easington Church derived some income from the marshes, and some from " le Hawenne," *i.e.*, the haven. As this haven is not given in Holinshed's list of creeks, etc., referred to elsewhere, it had apparently already disappeared in the sixteenth century.

In 1771 the church was 1,056 yards from the cliff. When measured in 1833 by Messrs. Hadfield and Pears, the distance was 968 yards. In 1882 it was 850 yards only, an annual loss of nearly 2 yards during 111 years.

In 1831 John Field surveyed " Ten Chain Close," Easington, which had a frontage to the sea of about half-a-mile, and it was found that in the 61 years that had

Photo by]   E ASINGTON TITHE BARN.   [W. S. Parrish.

elapsed since the enclosure of 1770, a strip 127 yards wide had been washed away, a yearly average of over 2 yards.

Only so recently as 1911 Firtholme House Farm at Easington, consisting of the farmhouse and buildings, and 130 acres of land, was sold for £650, whereas the mortgagees had lent £4000 upon it some time ago. This forcibly illustrates the deterioration of property in these parts, partly by floods and partly by erosion.

Mr. Matthews estimates the erosion here " during the past few years " at 5 yards per annum. In the period 1852-1888 it had been at the average rate of 12 feet per annum between Easington Lane end and Kilnsea ; and on surveying this stretch in 1898 Mr. Butterfield found that the average annual loss was 10 feet. Between

EASINGTON HALL, FORMERLY THE RESIDENCE OF THE OVERTONS, AS IT WAS IN 1770.

1852 and 1886 the loss had been 107 yards ; and between 1886 and 1898, a further 33 yards, or a total of 140 yards in 46 years.*

Mount Pleasant Cottage, built in 1876, bears a stone to the effect that it was then 616 yards from the sea. The lettering on this stone is now almost obliterated. The house referred to was formerly occupied by the late Dr. H. B. Hewetson. It is now known as The Tower, and is occupied by Mr. R. W. Walker, who informs

* " Naturalist," November 1904, p. 326.

OLD COTTAGE AT EASINGTON, PARTLY BUILT OF MUD.

OLD COTTAGE AT EASINGTON.

me that the distance in February, 1912, was 470 yards ; a loss of 176 yards in 36 years, or over 4 yards a year.

To-day Easington has many places of interest, including a fine old aisled tithe-barn and several old cottages.

Northorp township has disappeared. It was formerly within the parish of Easington. With Hoton, it is referred to in a surrender of William De-la-Pole to Edward

VIEW OF EASINGTON TO-DAY.

The Old Hall was situated where the shops now are, on the left of the photograph.

III. In the " Liber Melsæ," Northorp is called a manor, and according to the same authority " perished with Hoton, and was all gone in 1396." According to the History of Withernsea a " Northropp " is referred to in a document dated 1667 ; but this can hardly be the same place.

Another township, Hoton, was once within Easington, and had disappeared in the fourteenth century. (See also under Northorp).

Another place, Turmarr, once within Easington, had also disappeared in the fourteenth century. A field north of Easington, where there is a depression in the cliffs, is still known locally as Turmarr Bottoms.

CHAPTER XII

The Romans at Easington—Pottery from Site of Occupation—Oyster Shells, Coins, Etc.

THE discovery of the Roman remains referred to in the following note is some indication of the probability. elsewhere expressed, that there was once a Roman road along the coast of Holderness, the site of which is now seaward. It is evidence also of this locality having been occupied for a considerable time. Not only are there many Roman remains recorded at Spurn ; but the late Dr. H. B. Hewetson, with the late J. R. Mortimer, opened a number of British barrows at Easington and near Kilnsea, the crude earthenware vases, etc., from which were disposed of at Dr. Hewetson's death, and I regret that their present whereabouts are unknown. These tumuli have since been washed away, though now and again their sites can be traced on the peat near the beacon. About the same time as Dr. Hewetson carried on his excavations, I secured a human skeleton, which had been buried with a flint implement, in the peat, and this I still possess.

The place where the Roman remains referred to below were obtained, is now far out to sea.

The discovery was made at Easington in 1875, in which year Mr. W. Stevenson was staying there, and

noticed sections of two trenches filled in with dark
earth, in the cliffs. One of these occurred each side
of the road at Easington Lane end, as shown on the map
below. Each trench was 6 feet wide and 6 feet deep,
the cliffs at that time being 10 feet in height from the
beach. In the northern hollow, which appeared to be
the end of a dyke running for a considerable distance
inland, a few bones, oyster shells, and pieces of earthen-

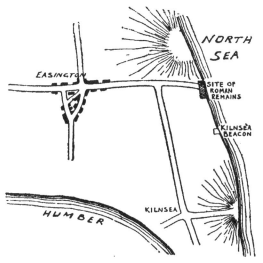

SKETCH-MAP SHOWING THE SITE OF THE ROMAN
REMAINS FOUND AT EASINGTON.

ware were noticed, which induced Mr. Stevenson to
examine the vicinity in some detail. Mixed up among
the débris of boulders, soil, etc., were some pieces of
pottery which Mr. Stevenson pieced together and
restored.

An examination of the vessels reveals the fact that
they were mostly used for domestic purposes, and
they also varied considerably in texture and in the quality

of the clay of which they were composed. The vessels are usually quite plain, though one fragment has the characteristic zigzag pattern marked upon it. They are roughly of two forms—the ordinary urns or vases, and flat dishes or basins. Of the latter type there are three examples, which only needed a little restoration to make them perfect. Fragments of other vessels of a similar type, with sides about $2\frac{1}{2}$ inches high, and about 6 inches in width, were also found. Of the " urn " type are remains of six or seven vessels, some sufficiently complete to enable their original form to be ascertained.

All the pottery is of the light or dark grey colour, and no fragment of the red Samian ware appears to have been found.

The animal remains include bones of the short-horned ox (*Bos longifrons*) and wolf or dog.

Among the other objects obtained were two land-shells (*Helix*), and a large quantity of oyster shells. The latter are of interest, as they clearly indicate the manner in which the Romans opened the bivalves (page 123). It will be noticed that a notch has been nipped out of the centre of each valve.

While the particular locality is lost in which these remains have been found, there are still on the Humber side of Spurn some " kitchen middens " in which part of a bronze brooch, fragments of pottery, and other Roman remains have been found within the last few years. A vessel of somewhat unusual type was obtained in the peat bed near Kilnsea Beacon by Mr. Murray. This was not quite complete, but has been restored. Various fragments of earthenware have been found at the same place.

More recently the late J. W. Webster, of Easington, had been working among antiquities in his district, and secured several interesting examples, which are now exhibited at the Hull Museum. Chief among them is a vase, which was found in pieces, but has since been

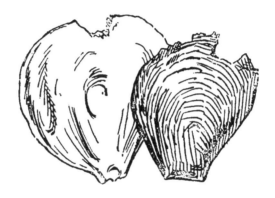

OYSTER SHELLS FROM THE ROMAN REFUSE HEAP AT EASINGTON.

Showing the method of opening by nipping a piece out of the front.

restored. It is 6 inches in height, 5½ inches across the top, and was found on the Humber shore in the locality where oyster shells are particularly numerous, and where fragments of pottery are common. Two or three other localities in the district yield Roman remains ; some of these are on the sea side. At one place six silver Roman coins were secured, one of which is attributable to Hadrian.

In addition to the remains from Easington and

Kilnsea, Roman coins have been found at Hollym, Withernsea, Hornsea, and Aldbrough. A hoard of over fifty Roman coins from Hollym is now in the Hull Museum. Roman pottery has also been found in fair quantity near the coastguard station at Aldbrough. Poulson (Vol. II., p. 523) refers to a beautiful figure of Mercury, and other Roman antiquities, having been found in the neighbourhood. There is no question that

ROMAN AND ROMANO-BRITISH EARTHENWARE.

The three large ones are from Easington and the two small ones from Kilnsea; now in the Museum at Hull.

many other objects of this character have been found, but have not been preserved, and therefore any evidence which they could have given is lost.

There is one interesting point in connection with these Roman remains found near Spurn. Some writers, in discussing the question as to whether Ptolemy's "Ocellum Promontorium" was Spurn or not, have raised the objection that Spurn has yielded no Roman remains. The finds just recorded are ample proof of Roman occupation ; and unquestionably more evidences have been washed away, which would have been the fate of the specimens found by Mr. Stevenson in 1875, had he not been there to rescue them.

CHAPTER XIII

DIMLINGTON : ITS BEACON—OUT NEWTON—RUINED
CHAPEL—HOLMPTON—MEASUREMENT OF LAND.

THE cliffs here are unusually high, and consequently
the rate of erosion is not so great as it would be if the
cliffs were lower ; though in recent years it has averaged
2 yards a year.*   Between 1771 and 1852 the loss was
1.8 yards per annum, or 125 yards in 81 years.   Between
1852 and 1881, 66 yards have gone, an average of 2.3
yards a year.

In 1851 the distance from Dimlington Farm and the
cliff edge was 250 yards.   It was 194 yards away in 1876.
When re-measured by Canon Maddock in 1898, the distance
was 83 yards, a loss of 167 yards in 47 years.   This gives
an average of over 3 yards a year at this point.   In
1901 half an acre fell in two tides.

The beacon on Dimlington was 48 yards away
when measured by Mr. Hatfield in 1833.   The beacon
and its site have now gone.

So long ago as 1346, a grant to the Abbot of Meaux
specified that the ancient manors of Saltehaugh, Tharlesthorp, Frismersk, Wythefleet, Dymelton [Dimlington],
and Rauenserodde had been reduced in value from £250

* Mr. Matthews estimates 3½ yards a year " during the past few
years."

126

to £20 per annum by the waters of the sea and the Humber;*

The old ruined " chapel," the prominent landmark

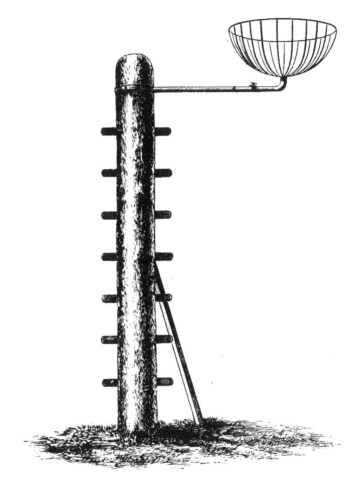

DIMLINGTON BEACON, NOW WASHED AWAY.

on this, the highest part of the Holderness Coast (125 feet high), was 147 yards from the cliff edge in 1833.* In 1882 it was 40 yards away (Reid) ; in 1898, 22 yards

* Thompson's " Ocellum Promontorium," 1824, p. 166.

(Maddock) ; in 1901, 20 yards ; and when I measured it in 1908, it was 12 yards from the cliff. In 1910 it was less than 10 yards.* Since then, only during the past winter, the last wall, which had stood the storms of six or seven centuries, has been pulled down—probably in order to repair a pigstye or mend a road.

Photo by]   *[Oxley Grabham.*

REMAINS OF THE OLD CHAPEL AT OUT NEWTON.

Pulled down in 1911 when within a few feet of the cliff edge.

Between 1771 and 1852 the average loss here was 0.8 yards a year only, or 65 yards in 81 years ; between 1852 and 1881 it averaged 3 yards a year, or 87 yards in 29 years ; and since that date some 30 yards have gone.

* "The Naturalist," 1908, pp. 342 and 384.

Mr. Matthews estimates the rate of erosion along these cliffs at 2 yards a year.

Between 1852 and 1876, at the " Gap," separating Out Newton Parish from Holmpton, the loss of land averaged over 3 yards a year.

The recent " History of Withernsea "* quotes a list of goods (a chalice, vestments, altar cloths, etc.), extracted from " Inventories of Church Goods, York, East Riding, 1552," as " the only scrap of evidence which the writers have met with relating to the old church at Out Newton."

In a Parliamentary Survey of the Benefices of the East Riding, made in 1650, we learn :—" There is a chapel at Outnewton belonging to the said parish [Easington] ; it is about two miles distant and is much decayed ; the hamlet fit to be annexed to Holmpton parish, being not a mile distant."

Poulson (1841) records that " between Out-Newton and Holmpton, there is a surprising appearance of a fresh-water deposit of marly clay on the top of the cliff, about 20 feet above high-water mark." The present position of the old lake bed at Out Newton is only a few feet above the beach, and I have known it as such during the past twenty years. The site of the old mere seems to be quickly disappearing, and the sections now are very poor indeed.

From many references to old documents in Poulson's " Holderness," pp. 379-80, it appears that Out Newton was a place of importance in early times. It is referred

* 1911, p. 90.

to in a document of the reign of Henry III., when the " Lord of Outnewton " is mentioned.*

At the time of the Domesday Survey there were, according to Boyle, 1280 acres in Holmpton. In 1800 there were 900 acres ; a loss of 380 acres between the years 1086 and 1800.

In 1786 the church is recorded by Tuke as being 1200 yards from the cliff ; in 1833 it was 1,130 yards ; in 1881 the distance was 1,042 yards, an annual loss of 1.6 yard. " Much of this loss, however, appears to have happened in late years ; the average since 1851 being 2.5 yards." † It would therefore seem that if in the latter thirty years there had been a loss of 75 yards, only 83 yards were washed away between 1786 and 1851.

In 1895, according to the British Association Report (Coast Erosion Committee) the distance was 1048 yards ; so that between 1881 and 1895 the loss has been practically nil ; the apparent increase in the distance is doubtless due to the difference in the exact place from which the measurement was made.

Just south of Holmpton, at " Old Hive," Pickwell gives the loss between 1802 and 1852 as 0.9 yards a year ; whereas between 1852 and 1876 it was 3.5 yards per annum, or 45 yards in the first fifty years, and 84 yards for the 24 years following! Reid adds :—" Measured on the old Ordnance Map the loss is very much greater, having been 5.5 yards annually since 1822. It is

* "Trans. East Riding Antiq. Soc.", Vol. IV., 1896, p. 56.
† Reid, in 1885.

difficult to reconcile these calculations ; and as the Ordnance Map is not very accurate [!] the real rate is probably nearest that given by Mr. Pickwell, though probably rather greater."

In 1898 the late Canon Maddock measured the " Cliff House Farm," from the barn to the edge of the cliff, to be 102 yards.

CHAPTER XIV

WITHERNSEA—SEATHORNE—FORMER CHURCH LOST— CREEK AT WITHERNSEA—PRESENT CHURCH ONCE IN RUINS — THE ROOS CARR IMAGES — BRITISH REMAINS.

OPPOSITE the village—or town—of Withernsea, the groynes, since their erection in 1870, have been helpful in preserving the sea front, and the promenade in more recent years has also assisted, as doubtless will the extension thereto just completed.

Just south of Withernsea " there is a sudden and great increase in the breadth of the strip of land lost since 1852," and the Geological Surveyors found the coast had so altered in this part of Holderness that they had the " new coast of 1881 " engraved for their map, instead of following the usual custom of putting their work on the old Ordnance Survey sheets.

For a mile and a half south of Withernsea, the annual loss between 1852 and 1885 was 3.3 yards, or a strip of land 100 yards wide in all! As this rate is much more than that covered by a longer period, the increased rate is attributed to the groynes which kept the beach to the north.

At " Nevilles," about two miles south of Withernsea, the rate of erosion was only 0.7 yards per annum between 1794 and 1852 (Pickwell) whereas between 1852 and 1882 it averaged 2.3 yards. In 1763 Neville's farm contained 140 acres ; and one " close," adjoining the sea, was 10 chains long. In 1845 it was 6 chains long, a loss of 144 yards in 82 years, or 1 yard 2 feet a year. In 1911 the measurement had reduced from 7 chains to barely 3 chains, so that between 1845 and 1911 a further strip, 98 yards wide, has been washed away.*

VIEW OF WITHERNSEA.

Shewing the old Queen's Hotel and ruined Church.

(From an old Print in the possession of Mr. W. Sykes).

The parish of Withernsea at the Inclosure in 1794 contained 800 acres ; in 1890 there were 745 acres, so that 55 acres had disappeared in about a century.

" Seathorne (*i.e.,* here meant as Withernsea) church, now in ruins," in 1833, was measured by Mr. Hatfield as

* " History of Withernsea,' p. 266.

417½ yards from the cliff. In 1895 it was 280 yards

WITHERNSEA, *circa* 1880

This view shows the old Pier, which, with the road in front of the picture, has been washed away by the sea.

(*After a Painting by G. Cammidge*).

away, according to the British Association Report ; a loss of 137½ yards in 62 years. The middle of the road,

" opposite the ruin " (now Withernsea's main street) was 278 yards away, and the Intake Farm House (west end) was 312 yards away. The Intáke Farm was only 134 yards distant in February, 1912, so that 178 yards have disappeared since 1833.

Prior to the groynes at Withernsea being built,

WITHERNSEA CHURCH IN 1829.

(*From Allen's " Yorkshire "*).

the erosion there was very severe. " In 1852 the distance of the southernmost house of the old village on the east side of the high road from the cliff edge was, in a line due east, exactly 600 yards. In 1876 it was 455 yards, so that the loss of land had been 145 yards in 24 years, or at the rate of 6 yards per annum. In 1652 the old poorhouses in what is now South Cliff Road, measured from the east end 425 yards to the cliff ; but in 1876 the distance had been reduced to 290 yards, a loss of 135 yards in 24 years, or 5·6 yards per annum. · · · Near the chapel-of-ease on the west side of the high road, the loss from 1812 to 1852 was 2.2 yards per annum ; then up to 1871 it was 3.3 yards per annum. In 1871 the distance from the high road at the corner of

WITHERNSEA CHURCH IN 1840.
(*After Poulson*).

the farm now occupied by Mr. Needler was 30 yards ; the whole of which has gone, including the high road itself. . . . Now the majority of the groynes are so wrecked and damaged as to be almost worthless, the erosion is steadily re-commencing, and year by year getting worse. During the last three years (1897-1900), over 12 yards of land in breadth have eroded beyond Queen's Terrace, southwards ; and in 1897, during a heavy and long-continued gale, the foundations of the sea wall were bared and the beach swept out to a depth unequalled for five and twenty years or more."*

Early in the fifteenth century the churchyard at Withernsea, having been washed away, an enquiry was held, and it was decided (in 1444) to rebuild the church on Priest Hill. In 1488 the church was completed and conse-crated.† In the time of Henry VIII. the church was "much decayed," and in 1609 was damaged by a storm, and was prac-tically a ruin until half-a-century ago.

Mr. G. Miles con-siders that the site of the old church of St. Mary, Withernsea (which was removed

THE INTERIOR OF WITHERNSEA
CHURCH IN 1841.
(*Poulson*).

to its present site on "Priest Hill" towards the end of the fifteenth century), is "at a point now covered by the sea, and about a mile due east from the corner of the inn now known as the 'Commercial.' Around this old church of course would be the old Withernsea, the 'Witforness'

* Mr. Cheverton Brown, in the *East Riding Telegraph*, April 7th, 1900.
† Mr. Matthews ("Proc. Inst. Civ. Eng.", 1905) states that Withernsea Church was twice moved, but I can find no evidence of this.

137

of Domesday, which held its fairs, etc., and 'what of local importance survived, was transferred to Hollym.' "*

Similarly, in a deed dated 1884, drawn up in connection with the Withernsea Estate Co., for land to the south of the remnants of the pier, we find " Mill Field, heretofore said to contain 7 acres 1 rood 20 poles ; but by reason of the encroachment of the sea, found to contain only 4 acres 3 roods 21 poles." Another plot " said to

WITHERNSEA PIER, *circa* 1880, NOW WASHED AWAY.

contain " 9 acres, only contained 5, and a further lot " said to contain " 4 acres, only contained 2.

As showing the former nature of the county in Holderness, or as proof of the accuracy of many of the old maps which indicate streams or rivers connecting the North Sea with the Humber, Holinshed's Chronicle, dated about the middle of the fourteenth century † says " Being come

* " History of Withernsea," p. 266.
† See " Ocellum Promontorium," pp. 202-3. Holinshed died about 1580.

about the Spurnehead, I meete yer long with a *riuer*
that riseth short of Withersie, and goeth by Fodringham
and Wisted [Winestead], from thence to another that
commeth by Rosse [Roos], Halsham, Carmingham
[? Keyingham], then to the third, which riseth above
Humbleton, and goeth to Easterwijc [Easternwick],
Heddon, and so into the Humber. The fourth springeth

VIEW OF SUTTON CHURCH.

The view which evidently " inspired " the artist who drew
the alleged view of Ravenser (see page 92).

(After Poulson).

short of Sprotleie, goeth by Witton [Wyton], and falleth
into the water of Humber at Merflete [Marfleet] as I
heare."

Possibly also the seaward end of this stream may
have served as a harbour, as in Holinshed's " List of such
ports and creeks, as our sea-faring men doo note for their

benefit upon the coasts of England," Sister-kirke, *i.e.*, Withernsea, is given.

As evidence of the probable existence of a navigable creek inland, from Withernsea, is the discovery of the

THE PRE-VIKING IMAGES FROM ROOS ·CARRS.

well-known " Roos Carr Images."* " Some labourers, who were employed in clearing out a dyke or ditch which had been made some years previously . . . in Roos Carrs, discovered, about 6 feet below the surface, in a

* See " Hull Museum Publication," No. 4, 1901.

bed of blue clay, a group of figures."* These were carved in wood, had quartz pebbles for eyes, and were inserted in a serpent-shaped boat. In Andrew Jukes' "Guide to the Museum of the Literary and Philosophical Society, Hull" (1860), it is stated this model was found "in clearing out what is now a dyke . . . but which appears formerly to have been a creek or haven connected with the Humber." The models are the work of pre-Viking Northmen, and are the earliest evidences of the settlement in this country of people from Scandinavia. Quite possibly their ship was wrecked or stranded, and their "gods," which were probably placed in the prow of the vessel, were lost or buried. Anyway, this forms a little more evidence of a creek, such as is shown in Lord Burleigh's chart, having been here much earlier. This chart also shows that between Withernsea and "Thorne" † there was a small bay, and another between Thorne and Tunstall. These are described as "Twoo small crekes for landing of fysher bootes [boats] wherin small shippes at spryng tyde may also entre and do annoyaunce."

In the Parliamentary Survey of the Benefices of the East Riding made in 1650,‡ we learn "the church of Withernesey is very much decayed, insoemuch that it

* They were found in 1836.
† Doubtless "Owthorne" is meant, although "Sisterkirke" is also shewn, a little distance inland, but this is as might be expected, as we are distinctly told that the "plotte" is made for the description of the Humber and the coast "wherfore though sum [some] hamlettes and villages of Holderness be left oute. It is not material for this purpose this was made."
‡ Trans. East Riding Antiq. Soc., vol. 4, 1906, p. 57.

is thought 300 li. will not repayre it ; the towne being much wasted by sea. That part which remayneth is near adjacent to Owthorne, and some parts near Ollime " [Hollym].

A fine neolithic polished flint axe-head was washed

NEOLITHIC FLINT AXE-HEAD, WASHED FROM THE CLIFFS AT WITHERNSEA.

from the cliffs at Withernsea many years ago, and is now in the Hull Museum.

Seathorne, in Domesday Book, is Outhorne or Owthorne, the " Sister " of Withernsea, the church of which has long since gone.

CHAPTER XV

OWTHORNE is sometimes referred to as Seathorne, and in
Domesday also as Torne. Mr. Reid (1885) pointed out
that since the Ordnance Map of 1852 was made, the road
leading to Waxholme from Owthorne has been washed
away, necessitating vehicles going through a farmyard.
During the past few years the path made along the cliff
edge for foot passengers, outside this farmyard, has gone,
as well as part of the farm buildings.

Between 1812 and 1885, 2 yards a year, or 146 yards
in all, have been carried away at this point.

A little nearer Withernsea the loss between 1786 and
1870 has been 168 yards, or 2 yards a year. The groynes
erected in 1870 were successful in staying the erosion.
As Reid points out :—" Even since 1822, the date of the
old Ordnance Survey, the village of Owthorne, with a
church and twelve houses, has been entirely swept away,
and Owthorne and Withernsea meres have both disappeared."

In 1833 the main post of Owthorne Mill was 833 yards
from the cliff.

The steeple of Owthorne Church was 22 yards from

the cliff in 1805, 8 yards away in 1814, and it fell in

OWTHORNE CHURCH IN THE YEAR 1800.
(*From Thompson's " Ocellum Promontorium "*).

1816 ; its site in 1833 being 18 yards seaward from the cliff edge.

In an old document quoted in the " History of

144

Withernsea,"* and dated 1741, reference is made to " A close called Dent Close, lying east of the Mar [mere] and west of Munge Close or Munge Croft in Owthorne." As half the mere has been washed away (the remaining portion being the present so-called " Valley Gardens "), it is clear that both Munge Close and Dent Close have gone since 1741.

OWTHORNE CHURCH IN 1806, FROM THE SEA.

(After Poulson, who probably based his drawing upon the engraving in "Ocellum Promontorium ").

Similarly, " Constable Field," at Owthorne, about 1800, measured over 11 acres ; in 1907 it was re-measured and found to be only slightly over 5 acres, a loss of 6 acres in less than a century.

John Phillips records (" Geol. of Yorks.," 1875, p. 73) that on his first visit to Owthorne, in 1826, two grave-

* 1911, p. 266.

stones remained in the western churchyard ; on one was an inscription containing the line " I must lie here till Christ appear." Two years later he saw the " bones of former generations " being washed out of the cliffs. In 1872 " the whole is changed to a promenade, which extends across and conceals the lacustrine deposit."

In 1786 Owthorne churchyard was reached by the sea, the chancel then being only 12 yards from the cliff

OWTHORNE CHURCH AT THE CLIFF EDGE IN 1797.
(*Poulson*).

edge. In 1793 the chancel was taken down, and six years later a faculty was obtained to take down the church.

The parish registers, etc., were removed to Rimswell Church, and contain a record in 1800 of " fifteen days' attendance at the old churchyard ' leading ' bones to the new churchyard " ; and there is also reference to a charge for " burying bones from the sand."*

* " History of Withernsea," p. 44.

Between Owthorne and Tunstall there was a small creek in the sixteenth century (see Lord Burleigh's chart, p. 209) which, as in the case of the one a little to the south, was " for landing of fysher bootes [boats], wherein small shippes at Spryng tyde may also entre and do annoyaunce."

The plan on page 149, the original of which is now in Rimswell Church, prepared about a century ago by Robert Stickney (the surveyor appointed by the Owthorne Inclosure Act of 1806), well shows to what a large extent the sea is responsible for the loss of the towns on the coast. From the dotted line representing the present cliff-edge, it will be seen that within half a century the church and churchyard, vicarage and grounds, houses, streets, drains and fields, have all gone ; and the farm shown at the top of the plan is now quite on the cliff edge, part of the buildings having been washed away in recent years.

A glance at this plan also shows how very much Withernsea has altered in the same time. On what is now Withernsea's main street (Queen Street), with its shops and inns, there was then not a single building.

Referring to Owthorne, Poulson, writing in 1841, says :

" A few years since, before the sea engulfed the last relict [*sic*] of Owthorne Church, a more touching and interesting spectacle could scarcely be witnessed by a reflecting mind than these " Sister Churches." Owthorne Church, standing like a solitary beacon on the verge of the cliff, perpetually undermined by the billows of the ocean, and offering a powerless resistance to their encroachments. The churchyard, and its slumbering inmates, removed from time to time down the cliff by the force of the tempest ; whitened bones projecting from the cliff, and gradually drawn away by the successful lashing of the waves ; and after a fearful storm, old persons tottering on the verge of life, have been seen slowly moving forth and recognising [!] on the shore the remains of those whom in early life they had known and revered. The old church still remained ; but the wide fissures in the walls, and the shattered buttresses, plainly told it must soon fall in the common wreck.

In 1786 the sea began to waste the foundations of the churchyard. In 1787 there were two bells in the tower, and the third broken. In 1796 the church was dismantled ; and in 1816, after an awful storm of unusual violence, the waves having undermined the foundations, a large part of the eastern end of the church fell with an awful crash, and was washed down the cliff into the sea ; many coffins* and bodies in various states of preservation were dislodged from their gloomy repositories, and strewn upon the shore in frightful disorder. Amongst the rest one coffin in particular . . . . These relics of departed greatness found a new place of sepulture in Rimswell. In 1822 the chancel, nave, and part of the tower were gone. In 1838 there was scarcely a remnant of the churchyard left."

A valuable relic is recorded from Owthorne, the present whereabouts of which it would be interesting to know. " On Thursday, November 8th, 1785, a canoe was dis-

* It is hinted " that during the washing away of the chancel of the old church, the coffin of a former rector was exposed, and the rector and clerk of that time *fought* for the ownership of the lead coffin ! " and " a skull, which projected from the cliff of Owthorne burial-ground was observed to be occupied by a robin red-breast, where she, undisturbed, built her nest, and reared her brood." (" Churches of Holderness," 1837).

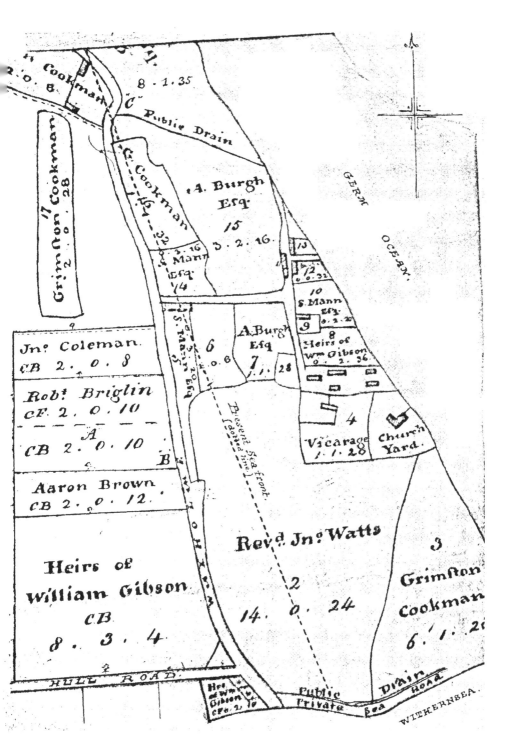

PLAN OF WITHERNSEA A CENTURY AGO, SHOWING THE LOST LAND.

covered in the clay, at the distance of about 50 yards south-east of the church, perfectly entire, with a broad stern, 12 feet in length and 4 feet broad. Two or three tides preceding the above discovery were extremely high, and set very hard upon the shore opposite the church. The shore, being for many years previously a fine sand, which was totally removed by the action of these violent tides, and a blue clay appeared, upon which were prints of birds' feet, particularly swans, which are supposed to have been imprinted on the clay centuries ago, no swans having been noticed upon this coast within the last hundred years. An old man, says the narrator, remembered a canoe being found about sixty years before, but of much less size."*

* Poulson, vol. 2, pp. 407-8

CHAPTER XVI

NEWSHAM—IMPORTANT IN DOMESDAY TIMES—WAX-
HOLME—SAND-LE-MERE : ITS CONNECTION WITH THE
HUMBER—TUNSTALL—HILSTON—HILSTON MOUNT.

NEWSHAM or Newsom, formerly within the parish of
Owthorne, seems to be entirely lost in the sea, and very
little information in reference thereto is available. · The
place is referred to in a deed dated 1662. In Domesday
times it was apparently a place of some importance, as
" In Niuuehusum Ernuin had five carucates of land
and two oxgangs to be taxed," said to be equal to at
least 600 acres, " where there may be five ploughs.
Drogo has now there one sokeman, and nine villanes and
seven bordars, with three ploughs and twenty acres of
meadow ; one mile long and one broad. Value in King
Edward's time sixty shillings, now forty shillings."
" There was a chapel here also, as well as at Waxham
and Rimswell, and in the 18th year of Richard II. was
conveyed to Kirkstall, with so many others in Holderness ; neither its site nor any other vestige remains."*

In 1783 a writer from Withernsea, to Dade, the historian, stated " no person in the neighbourhood can give
me any information of such a hamlet as Newsham."

At Waxholme, near Withernsea, between 1844 and

---

* For other early references to Newsham see Poulson's " History
of Holderness," vol. ii., pp. 313-4.

1852, the loss was 1.2 yards a year, and between 1852 and 1881 it was 1.4 yards a year, thus 45.5 yards were washed away in 37 years.

At the most northern farm in Waxholme the loss between 1844 and 1852 was 2.6 yards a year, whereas between 1852 and 1885 it has averaged 1 yard per annum only. A little further south the rate of erosion was 1.3 yards between 1855 and 1885.

ALL THAT IS LEFT OF WAXHOLME.
The road leading from Withernsea is washed away.

Speed's map of East Yorkshire (1610) shows a stream joining a mere near Waxholme with the Humber.

Waxholme may practically be said to have now disappeared. In Domesday times in Washam, as it was styled, " Torchil and Tor had two carucates of land to be taxed, and there may be two ploughs there. Alelm has now there of Drogo six villanes and four bordars, with two ploughs and sixteen acres of meadow. One mile long and four quarentens broad. Value in King Edward's time twenty shillings, now ten shillings. Six oxgangs are returned as soke to Withernsea."

There are many references to " Waxham " in documents of the fourteenth, fifteenth, and sixteenth centuries.

A chapel formerly existed here, and in 1394 it was

THE COTTAGES AT SAND-LE-MARR, NOW AT THE CLIFF EDGE.

given to Kirkstall by the Abbot and Convent of Albemarle. In the reign of William and Mary the chapel is referred to as being in a dilapidated state ; it has since gone.

In 1833 the " Preventive Watch-house " erected in 1800 near Sand-le-Mere was measured by Mr. Hatfield and found to be 84 yards from the cliff. The building was later occupied by the coastguards, and in February

1912, the corner of the northernmost building was within 4 feet of the cliff edge. Part of it has since gone. Thus over a yard a year has gone during the past 80 years.

Of Sand-le-Mere, Poulson says but little :—" Sandley Marr is now (1841) the site of a poor cottage on the cliff, one mile from Tunstall, and is destitute of all attraction except the green luxuriance of broad acres, and the wide and solitary expanse of the German Ocean. The beach affords excellent materials for the repairs of the Holderness roads " !

The bank erected across the seaward end of the hollow, once the Mere, is frequently wrecked by the sea, and has to be renewed. This occurred as recently as 1911.

In 1910, Mr. G. W. Lamplugh, F.R.S., described a find of estuarine shells in the alluvial hollow at Sand-le-Mere, which the recent storms had well exposed in the clays and marls on the beach.* It was the first occasion upon which estuarine shells had been noted in the clays of the Holderness coast. The shells consisted of the cockle, mussel, *Scrobicularia*, *Tellina balthica*, and *Hydrobia*. " The deposit and its fauna clearly indicate a quiet estuarine creek regularly invaded by the salt water tide, with sea-level approximately the same as at present. . . . From the present position of the shore-line in relation to the hollow, it might seem at first sight certain that the salt-water of the estuarine stage flowed in at its eastward end from the open sea. Yet it is more probable

* " The Naturalist," January 1910, p. 11.

that the inflow was from the opposite direction from the Humber, by way of the gap at Keyingham. The fauna is essentially that of the Humber muds, and the sediments are such as one should expect to find in a creek of the Humber. . . . I surmise, therefore, that the estuarine beds of Sand-le-Mere were accumulated when the hollow was a blind inlet of the Humber, which has now been decapitated by the recession of the coastline."

The discovery of the Roos Carr Images has some bearing on this question (see page 140).

In Saxton's map of 1577, and also in later maps which are probably based on Saxton's, a distinct mere is shown just north of Waxholme ; and connecting this mere with the Humber, between Keyingham and End-holm, a well defined stream is indicated. This gives support to Mr. Lamplugh's supposition.

In 1898 I found the remains of a lake-dwelling at Sand-le-Mere, when the peat bed was well exposed on the beach ; in that part of the section there was no evidence of estuarine conditions. These appear to have since disappeared by the action of the sea. A year ago I was able to confirm Mr. Lamplugh's find of estuarine shells, and obtained a number of specimens. There appear to be two distinct sets of strata, however, in one of which there are proofs of fresh-water conditions only, including remains of the pike.

In the time of the Domesday Survey " In Tunestale there were 7 carucates in soke to Chilnesse [Kilnsea], and one carucate in soke to Witforness [Withernsea]."

This is estimated as 1280 acres. At the enclosure of 1777 there were 800 acres ; 480 acres having been washed away.

In 1786 the Church at Tunstall was 924 yards from the cliff. In 1833 it was 763 yards. In 1881 the distance was reduced to 733 yards, a loss of 191 yards in 95 years, and an average of 2 yards a year. In 1832 Mr. Hatfield measured the distance as 763 yards ; and in 1853 it was 737 yards away. According to the British Association Report, the distance in 1895 was reduced to 691·6 yards.

In 1833 the distance from " the middle of the road at Tunstall Nook " to the cliff, measured by Mr. Hatfield, was 214 yards. Poulson records that by the sea " 100 acres are gone during the last 60 years " (1780-1840).

Monkwike, once within the manor of Tunstal, has entirely disappeared. In the times of the Domesday it was evidently of importance, as we read :—" In Moncuuic two carucates of land to be taxed. Land to two ploughs. Six villanes have there three ploughs, and they pay ten shillings."

In 17 Edward III. Wm. Ross de Hamlake held one wind-mill and divers free rentals in Monkwike, of the Provost of Beverley. Other early references to Monkwike are given by Poulson, who records that in his time (1841) " the manor lies along the sea-cliff. . . . It has suffered materially from the devastations of the sea, and not many years hence will be entirely gone " ; a prophecy which has since been fulfilled.

Hilston is called **Heldoveston** in the Domesday Survey,

where we learn that " In Heldoveston and Hostewic (Owstwick) Murdoc has seven carucates of land to be taxed, and there may be seven ploughs there. Drogo now has it and it is waste. Valued in King Edward's time fifty-five shillings." In a charter of 1272 it is called Hildofston ; later, in the reign of Richard II., " Hildeston," which is a near approach to the present name.

Hilston Church was rebuilt in quite recent years ; though the doorways, etc., from the old building, have been preserved. The old church was " one of the smallest parish churches in Holderness ; being only nineteen paces long by six or seven paces broad."

Between 1777 and 1852 the average rate was 1.1 yards per annum, opposite the church, and between 1852 and 1885 the rate kept the same ; thus 119 yards have gone in 108 years. According to the British Association Report for 1895, the church was 1056·6 yards from the nearest point to the cliff edge, in that year.

In 1832 Hilston Mount was 1200½ yards from the cliff, according to Mr. Hatfield's measurements, and in 1833 the " middle of the road at Whale Nook " was 209 yards away.

CHAPTER XVII

G<small>RIMSTON</small> G<small>ARTH</small>—M<small>ONKWELL</small>—R<small>INGBOROUGH</small>—A<small>LD</small>-
<small>BOROUGH AND ITS</small> D<small>ESTROYED</small> C<small>HURCH</small>—R<small>OMAN</small>
R<small>EMAINS</small>—C<small>OLDEN</small> P<small>ARVA</small>—G<small>REAT</small> C<small>OLDEN</small>—M<small>AP</small>-
<small>PLETON</small>—M<small>EASUREMENTS SINCE</small> 1786—R<small>OWLSTON</small>.

T<small>HE</small> cliffs at Grimston are over 70 feet high. Between 1883
and 1885 the loss was 1.4 yards on an average, or 65 yards
in the 52 years. On the former date the house on the
site of Grimston Old Hall was 325½ yards from the cliff,
and at the same time the new Hall was 725 yards away.
In February 1912, Mr. J. Atkinson kindly re-measured
these for me, and reports that the Old Hall is now 239½
yards from the cliff, and the new Hall is 652 yards. At
the first point, therefore, 86 yards, and in the second,
73 yards, have disappeared since 1833.

During the Domesday Survey, Grimston contained
six carucates of arable land, of which four were enclosed
as a soke appendant to the manor of Withernsea ; and
the other two as forming one of the berewicks belonging
to St. John de Beverley, subject to the danegeld. Here
was also a waste. Obviously there have been considerable changes here since Norman times.

Recently an unusually finely-made neolithic axe-
head was found in the cliffs near Grimston Garth. It is
particularly well formed, and of an unusual type.

Monkwell, a village once near Ringborough, appears
to have practically all gone.

In Domesday, " Ringborough was considered jointly with Garton as a soke of Esington. One carucate is

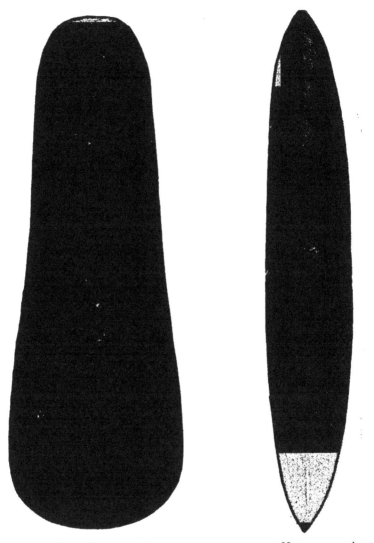

also returned as a soke to Aldborough, and one to Kiln-sea." Later, in the fourteenth century, " John Ros de

Ringburgh " had a wind-mill, eighteen tofts, seventeen bovates of land, etc., in Ringborough.

In Poulson's time the place was " reduced to a single farmhouse and farm," and, of course, much has gone since then.

Between 1833 and 1885 the average loss here was 1.6 yards, or 61 yards in all. The rate of erosion has varied during that time, and towards 1885 was increasing. In 1833 the west end of Ringborough Farm House was 305 yards from the cliff.

" During the past 24 years " (Matthews, in 1905) " the encroachment has been about 70 yards, or on an average about 3 yards per annum. The cliffs here rise to 50 feet in height."

The ancient township of Aldborough was certainly at some distance to the east of the present village, and its site is now beneath the waters of the North Sea. As is the case of other Holderness villages, it has travelled westwards as the sea has carried away its foundations. In the present church are relics from a previously existing Saxon edifice, doubtless the one now washed away. The principal item is a fine circular sundial, now built into the wall, upside down, and inside the church. The inscription reads VLF HET ÆRIERAN CYRICE FOR HANVM 7 FOR GVN-WARA = " Ulf bade rear a church for the poor (or for himself) and for the soul of Gunware."

The cliffs here are fairly high (over 60 feet). In 1786 the present church was 2,044 yards from the cliff. In 1832 the distance was 1,953 yards, as measured by Mr.

Hatfield. In 1882 the distance was 1,880 yards, a loss of 164 yards in 96 years, or an average of 1.7 yards a year. In 1895, according to the British Association Report, the distance was 1868 yards, The " Beer-house near the sea " was measured to the cliff in 1832 by Mr. Hatfield, and found to be 153½ yards away. If this is the same as measured by Mr. Petch* (see below) the present distance is 30 yards! That is, a loss of 123½ yards in 79 years. The old wind-mill at Aldborough was 1769 yards from the cliff in 1895 (see British Association Report for that year).

From measurements made by Mr. Petch at four points at Aldborough between 1893 and 1911 † it seems that 28½ and 11¼ yards, and 22¼ yards respectively were washed away in the eighteen years. At another part 20 yards were washed away in ten years. As these four points are less than a mile apart, it will be seen how much the rate of erosion has varied in the short distance. On the average, Mr. Matthews estimates the erosion at Aldborough now to be from 2 to 2½ yards a year.

The site of the original church at Aldborough, i.e., the church which contained the Saxon inscribed stone which is now let into a wall in the present structure, is far out to sea.

A quantity of Roman pottery, etc., has been taken from the cliffs in recent years.

Colden Parva or Little Colden had a chapel dedicated to St. John the Evangelist, and was conveyed to Kirkstall

* Mr. Petch measures from the south-east corner of the permanent buildings of the public-house." † " The Naturalist," November 1911.

in the reign of Richard II. Poulson (1841) records that
" this chapel, with a portion of the village, has suffered
from the devastations of the sea ; it was swept away
about a century and a half since," *i.e.*, about 1690. Then
we learn that " the living exists though the chapel has
been destroyed."

Poulson gives details of some surveys taken towards

Photo by]　　　　　　　　　　　　　　　　　　　　[*C. W. Mason.*
HUT AT COLDEN ORNAMENTED BY WRECKAGE.

the end of the eighteenth century, which show that in
less than ten years over 6 acres were washed away here :—

			Surveyed in 1788.			In 1787.			Lost.		
			A.	R.	P.	A.	R.	P.	A.	R.	P.
Sheep Field	..	..	29	0	10	27	0	33	1	3	17
North Field ..	..	..	13	3	4	10	0	0	3	3	4
North Close ..	..	..	28	3	29	27	1	15	0	3	10
Total	..	..	70	3	39*	64	2	8	6	1	31

* These figures are as given by Poulson.

Colden or Great Colden seems to have suffered severely. In the Domesday Survey (1086) there were 1920 acres in Colden. In the year 1800 there were 1100 acres —a loss of 820 acres in 714 years. Between 1764* and 1833 Pickwell estimated the annual loss 1.3 yards, or 92 yards. Between 1852 and 1885 it averaged 1.7 yards, or a further 38 yards, in addition to that lost between 1833 and 1852.

Reid (1885) records that "immediately north of Frank Hill, in a detached portion of Colden Parva parish, there has lately been a very great loss of land, no less than 115 yards having gone since the Ordnance Map was made [1852]; this gives an annual rate of 3.8 yards. Southward, as far as Aldborough, the average loss has been about 2 yards."

The "village" of Colden now consists of one or two farmhouses only.

Between 1786 and 1874 the loss at Mappleton was 2·3 yards a year, or 206 yards in all, and between 1852 (the date of the old Ordnance Map) and 1882, the loss varied from 40 to 50 yards in different parts of the parish.

In 1832, Mappleton Church was 507 yards away. At the present time, according to Mr. Matthews, the cliffs are being washed away at the rate of between two and three yards in a year.

The Rev. D. Hepburn Brown has kindly given me

* The distance from the nearest farm-house to the sea was 180 yards in 1764, and 90 yards in 1833, as measured by Mr. Hatfield

the following particulars extracted from the Mappleton Church Registers :—

" Distance from church to the sea cliff in a direct line :—

        April 20th, 1786........28 chains 76 links.
        October 1st, 1827 ......25 chains 45 links.
        September 23rd, 1835..25 chains 17 links.
        August 20th, 1847......22 chains 61 links.
        September 18th, 1849..22 chains 15 links.
        April 17th 1858........21 chains 62 links."

He also measured it again (February, 1912) and found the distance then was 418 yards, a loss of 2 chains 62 links since 1858.

In 1833 Mr. Hatfield measured the distance between Rowlston Hall and the cliff edge, and found it to be 867 yards. At the same time the iron gate near the lodge by the Hall was 556 yards away.

CHAPTER XVIII

HORNSEA BURTON is recorded by Poulson in 1840 as depopulated, though during the Domesday Survey it was
one of five sokes belonging to the manor of Hornsea,
and had two carucates of land under the plough. In
1200, Galfrid de Oyry granted to Fulk Saucey as much
land as was valued at 100 shillings yearly (a large
amount in those days), viz., 14 oxgangs out of his domain
in tillage, together with the pasture he held adjoining
the " Meer " at Hornsea.

Various grants of land in Hornsea Burton, made between 1228 and 1328, are quoted in the " History of
Holderness," p. 339.

According to " Kirby's Inquest " in the thirteenth
century, the heirs of Gilbert de Mapleton held in Hornsea
Burton " six carucates of land," say 720 acres ; in 1852

165

there were 409 acres, and now there are considerably less.

Whytehead, in 1786, recorded that Hornsea Burton continued to repair its own roads. " In Burton there are supposed to be 366 acres, in Southorpe 580. There is nothing now worth notice in the place, consisting only of one or two farm-houses."

Between 1845 and 1876 the loss of land at Hornsea Burton Farmhouse was 1·3 yards per annum, while between 1876 and 1882 the loss was five yards annually, the increase probably being due to the erection of groynes at Hornsea. Thus between 1845 and 1882 the loss was 71 yards. Opposite the brickyard at Hornsea Burton the average loss between 1845 and 1882 was 1·6 yards— a total of 60 yards. In 1882 the sea was only ten yards from the cottage.

At the present time the cliffs are being washed away here at the rate of $2\frac{1}{2}$ to $4\frac{1}{2}$ yards per annum.*

In his " list of such ports and creeks, as our sea-faring men doo note for their benefit upon the coasts of England," which was published about the middle of the sixteenth century, Holinshed mentions " Horneseie-becke," which would appear to confirm the information given on Lord Burleigh's chart reproduced on page 209.

About 1228 Walter de Spiney gave to Meaux Abbey his " whole profit of merchandise and of every ship applying at the port of Hornsea." It seems that Walter de Spiney's power of making such a grant was disputed,

* E. R. Matthews, *loc. cit.*

Photo by]   [*Payne Jennings.*

A VIEW OF HORNSEA, FROM THE PIER NOW WASHED AWAY.

Photo by]   [*Barr.*

THE OLD "PROMENADE" AT HORNSEA, AFTER THE HIGH TIDES OF MARCH, 1906.

the profits collected on vessels lying north of Hornsea Beck apparently belonging to the domain of Hornsea, while south of the stream they belonged to the lord paramount of Holderness. Consequently, the occupants of Meaux Abbey did not receive the benefit of these tolls.*

The pier at Hornsea is referred to in a petition of 1558. From an inquest held (7 James I.) it appears that this structure had cost £3000 (a very considerable sum in those days), and that 2500 tons of timber had been necessary to repair it.

A previous inquisition held at Hedon in 1400 showed that in 1334 Meaux Abbey held at Hornsea Burton 26 acres of arable land, for which they received 2s. per acre in rent, but of which at the close of the century about an acre remained. Evidently in 76 years the entire 25 acres had been washed away. Similarly, in 1609, an oath was made to the following effect :—

" We find decayed, by the flowing of the sea, in Hornsea Beck, since 1546, 38 houses, and as many closes adjoining. Also we find, since the same time,

* " But Walter did himself give unto us two locks in that same place, and all his toll and boardtoll at the sea, nothing being withheld, which pertained to him and his ancestors at the sea of Hornseabeck, on the clear understanding that all the aforesaid toll and boardtoll should be towards our hospitium, our own ship excepted, the toll from which he assigned to the needs of our convent. The toll even is on all goods and the boardtoll is on every ship mooring to the shore *four pence*. But, however, revenues of this kind in that place we never had, especially seeing that all revenues of this nature on the northern side of the stream, called the Beck, are known to belong to the lordship and liberty of the vil of Hornsea, and on the southern part of the said stream, on the shore of Burton, to the lordship of Holderness."—*Meaux Chronicle*.

Photo by| *[H. S. Harker.*

SOUTH CLIFF COTTAGE, HORNSEA, ABOUT 200 YARDS SOUTH OF THE
MERE STREAM.

The position of the old road can be seen between the house and the shed.

Photo by]   *[F. H. Wood.*

HORNSEA, SHOWING DELAPIDATED GROYNE.

The peat deposit is in the foreground.

decayed in ground the breadth of 12 score yards throughout the field of Hornsey, being a mile long, and parcel of the aforesaid manor. We further find that there will be great hurt and damage to the king's demesnes and pasture grounds near adjoining the said Hornsey Beck, within the manor of Hornsey, to the great hurt and impoverishing of the inhabitants of Hornsey, if that a present remedy be not made, either by re-edification of a peare or some other good defence for the same, for the safeguard of the said lands and country adjoining. And further, for the charge of the same, we find that the last peare built at Hornsey cost £3000 or thereabouts, and it will cost much more than it did then. . . . John Galloway, of Hornsey, pannierman, of the age of 80 years, says he had known 39 houses and 39 closes wasted away, of the yearly rent to the king of 58sh. 6½d., and that there doth usually every year waste the breadth of 40 feet, which is more than heretofore ; and that there are divers meadows and pasture grounds, called the King's Demesnes, of the yearly value of £11 18s. antient rent, which will in a short time be wasted and consumed, with a great part of the town of Hornsey, without a peare, which he thinketh will amount to 2500 trees. Edward Harrison, of Seeton, husbandman, aged eighty years, says that he has known 300 yards washed away, and that there was a peere at Hornsey Beck, during the continuance whereof the decay was very little."

On November 1st, 1757, Mr. Joseph Harrison measured " the distance from the north-east corner of Robin Maudley's house, at the seaside, to the edge of the cliff, along the balk, next the ditch, it was 61 yards 4 inches. April 2nd, 1759, . . . the distance was then 50 yards, so that in one year and five months the

Photo by]   *[H. S. Harker.*

THE LAST REMNANTS OF THE SAND-BANK SOUTH OF HORNSEA; 200 YARDS
SOUTH OF THE REMAINS OF THE PIER. FEBRUARY 1912.

Photo by]   *[H. S. Harker.*

NEARER VIEW OF THE REMAINS OF THE OLD SAND-BANK, HORNSEA.
FEBRUARY, 1912.

sea had gained 11 yards and 4 inches ; at the same time the distance from the beacon to the edge of the cliff was just 19 yards. The foundations of the house alluded to were washed away in 1785, and the beacon was removed about 14 years before that. In the year 1786 the distance from the church to the seaside was measured by Mr. John Tuke, surveyor, of York, when it was found to be due east 1113 yards. Mr. Harrison took the distance from the cliff in 1759. The distance from the church (east end), in December 1876, was only 1000 yards, making a deficit of 133 yards from the period of its ad-measurement by Mr. Tuke."

In a letter written by the rector of Atwick, dated September 19th, 1787, he states, " the place where the stream dyke empties itself into the sea for about eight months in the year, when there is a current from the Mere is . . . called the Beck ; near this beck the town was situated. Two or three years ago the Beck took another current to the sea 140 yards southward from the place where Robt. Maudley's house stood by the sea,* overflowing its banks, and filling up with sand its antient course, so that Mr. Bethell's manor is increasing in the same proportion as Mr. Constable's is decreasing. . . . Hornsea Beck has now altogether disappeared."

There was a bridge over Hornsey Beck in 1440.

In 1390, Robert Ticlot of Hornsea Beck willed to his wife Johan a ship called " *Fartoft*," in order that she

---

* The positions of both these streams are shewn on an old map in my possession, partly reproduced on page 178.

might make provision in the church of Hornsea for her own soul and the souls of her father and mother. He also left a small vessel, called " *Maudlin*," to his brother, and another to John Skelton for the same purpose. These early references indicate that vessels were safely harboured at Hornsea. It is also interesting to notice that the one private token issued in Hornsea by Benjamin

THE MARINE HOTEL, HORNSEA, IN 1845.

Much of the cliff shown in this copy of an old print has been washed away.

Rhodes, and bearing the date 1670, has a representation of a full-rigged ship on the obverse.

Even as early as 1257 Henry III. granted a charter to the abbots and monks of St. Mary's, York, for the holding of a market at Hornsea every Monday, which was continued until the end of the eighteenth century.

Up to the time of the dissolution of the monasteries by Henry VIII., Hornsea, " with its tithes, trade, and fisheries was the most valuable possession of the abbey."* In addition to the ships' tolls, the land tithes, wrecks, assize of bread and beer, and tolls from the markets, the abbot claimed a toll, known as "chiminage," from strangers

passing through the town. To assist the abbots in exercising their judicial functions there were gallows, tumbrils, pillory, and a prison, and writs were frequently issued against the abbots " by reason of their oppression and rapacious tyranny."

* Fretwell's Hornsea, 1894, p. 37.

CHAPTER XIX

Hornsea—Inquisition of 1607—Effect of Groynes—
Southorpe—Northorpe.

The rate of erosion at Hornsea is very irregular, and varies considerably over different periods. At the northern end of Cliff Lane it averaged 2·5 yards for 67 years, or 167·5 yards in that period, whereas near the Marine Hotel, where the cliffs are partly protected by groynes, the loss is only 1·9 yards a year, or 123 yards in the same period.

Pickwell quotes an " Inquisition held at Hornsea on the 28th of April, 1607, James I.", as under :—" We find decayed by the flowing of the sea in Hornsea Beck, since the first year of King Edward VI., 1546, thirty-eight houses, and as many little closes adjoining. Also we find, since the same time, decayed in ground the breadth of twelve score yards throughout the fields at Hornsea."

Mr. Reid says * " if the old inquisition held at Hornsea quoted by Pickwell, can be believed, the loss for the previous sixty-three years was 4 yards annually ; but the expression ' twelve score yards ' does not look like exact measurement, and the statement made by one of the witnesses that ' there doth usually every year waste the breadth of forty feet, which is more than hereto-

* Geol. of Holderness, p. 95.

fore,' seems scarcely trustworthy. No doubt the destruction of the pier would cause a great increase in the rate of denudation for a short time, and it might reach the large annual amounts here mentioned. The increase however, could only make up for the former protection of that part of the coast, the projecting portion being merely cut back till it attained the general line of cliffs on each side. The *average* rate of denudation for a larger period would remain almost unaffected. As far as can be made out, the loss at Hornsea has averaged a little more than two yards annually, but Mr. Bell* places it as high as three yards between 1786 and 1853."

In 1786 the distance between Hornsea Church and the cliff was 1133 yards, in 1832 (according to Poulson) it was 885 yards, a loss of 248 yards in 66 years, or nearly four yards a year during that period. According to the Ordnance Survey, the distance from the church to the nearest point of the cliff in 1895 was 898·3 yards.† From this it would appear that the 1832 measurement of 885 yards is not reliable, and therefore the estimate of four yards a year is exaggerated.

With regard to Hornsea Church, there is a tradition that when built it was 10 miles from the sea, and the following lines are said to have been inscribed on the steeple :—

> Hornsea steeple, when I built thee,
> Thou was 10 miles off Burlington,
> 10 miles off Beverley, and 10 miles off sea.‡

---

* " Rep. Brit. Assn.," 1853 p. 81.
† See " British Association Report," 1895.
‡ Poulson, 332.

THE SEA-WARD END OF THE MERE STREAM, OR "STREAM DYKE," HORNSEA.

THE REMAINS OF AN EMBANKMENT ERECTED AT HORNSEA ABOUT FIFTY
YEARS AGO BY THE LATE J. A. WADE.

M

There appears to be no evidence of either these measurements or of the inscription.

The erection of strong sea-defence works at Hornsea has practically resulted in erosion there being stayed,

PLAN OF HORNSEA, DATED 1801.

otherwise in time the well-known Mere would have been tapped by the sea, with disastrous results to Hornsea. Mr. Matthews states that within the memory of comparatively young people at Hornsea, " hotels, houses and cottages have had to be pulled down owing to the per-

Photo by]

HORNSEA PARADE IN 1907.

[R. Williamson.

sistent advance of the sea, and many have been swept away by the waves.'"

In the old chart (*temp*. Henry VIII. . . . see page 209), there are instructions to sailors, etc., and opposite the harbour and pier at Hornsea they were told to " warpe in and owte." The pier at Hornsea, shown on this old chart, was evidently destroyed before the year 1609.

It occasionally happens that the erosion of the coast is not altogether a disadvantage. For example, in 1770, the corpse of a murderer and smuggler named Pennel, was bound round with iron-hoops and hung on a gibbet on the north cliff, until such time as the " ornament " was washed away.

In 1786 the acreage of the lordship of Hornsea was estimated at 2013 ; to-day Mr. T. Hornsey informs me there are 2924 acres, so that either the former estimate is wrong, or the districts are not the same, as unquestionably much land has been washed away.

As already stated, Southorpe contained 580 acres in 1786. Now everything has been washed away. It was situated immediately south of Hornsea—hence the name —being the South Thorp, or village, with respect to that place. In Domesday times it contained a carucate and a half of arable land.

Northorpe village was formerly situated north of Hornsea, bearing a similar position on the north side to that of Southorp on the south. Little appears to be now known of it beyond the name. Poulson records that old people recollected seeing stones, etc., being dug up there, evidently parts of buildings.

Photo by] HORNSEA PARADE AS REPAIRED IN 1910. [R. Williamson.

CHAPTER XX

HORNSEA MERE has been the cause of much trouble in the past, and it is evident that the fish and fowl it harboured were worthy of consideration. So long ago as 1260, William, 11th Abbot of Meaux, claimed the right of fishing in the south part of the Mere, and this right was opposed by the Abbot of St. Mary's. It was decided to settle the controversy by combat. Champions were found on either side, and after a fight, which, it is stated, lasted all day, the fishery was relinquished by the Abbot of Meaux.

Poulson quotes " Burnsell on the East Riding," preserved in the British Museum, as follows :—

" A water prettie deep, and always fresh, about a mile and a half long, and half a mile broad, well stored with fish ; it hath in it three little plots, two of them full of egs of Tems [? Terns] at the season, and birds as can be imagined ; it is fed with the water that ren into it of the adjoining higher grounde from the north, south, and west, eastwards it runs up into the sea by a ditch, call'd the stream ditch, when the clew is open'd, there are mannie springs in it also ; the soyle is in some places gravel'd, in others a perfect weedy morass ; that this marr hath been through some earthquake and settling of the ground, with an overflow of water there-upon, seems probable, speciallie if that be true which I was

Photo by]

HORNSEA MERE.

[R, Fortune.

told that there hath been seen old trees floating uppon, and deca'd nutts found cast on the shore, but, however, that be this is certain, that in the sea cliffs against Hornsey, which is scarce a mile of, there is both wood and nuts to be found, and there is now or was lately there, at the downgate a win of wood which looks as black as if it had been burnt, which I think is occation'd through the saltness of the sea water overflowing it, which both preserves wood better than fresh water, and also by its saltness, and consequently greater hotness, help to turn it black, all this intimates that there hath been an inundation there; but when no historie, I believe, relates, unless it was in that earthquake, which was so generall through the world in the time when Valentinian and Valens were consulls, anno Christ 368; unless we should think as the vulgar say, those things hath been there ever since Noah's flood; this place bids as fare for anie other place I have herd of yet; I scarce think either wood or nutts can continue so long, though kept never so close from the violent motion of the air, &c."

Bearing on this, it is interesting to note that a section of the old mere was exposed in the cliffs in 1906, and a description of this, with a list of the shells, seeds, etc., then found, appeared in *The Naturalist* for that year (page 420).

One fact brought out in connection with this paper was that the numerous remains of animals and plants preserved in what was once the bed of an easterly extension of the present Mere, or more probably another mere altogether, clearly indicate that the mere had not been encroached upon by the waters of the North Sea. The fauna and flora is such that certainly could not

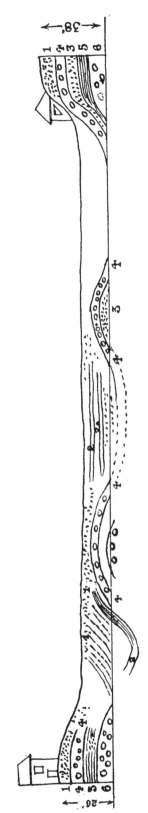

SECTION IN THE OLD MERE BED (450 FEET LONG) EXPOSED IN THE CLIFF, IMMEDIATELY NORTH OF THE MARINE HOTEL, HORNSEA, IN JUNE 1906.

1 = recent gravel ; 2 = marl with fresh-water shells ; 2a = peaty layers ; 3 = glacial gravels and sand ; 4 = tough boulder clay ; 5 = laminated glacial clays ; 6 = lower boulder clay.

have lived in water containing any appreciable admixture of salt.

On geological evidence, therefore, it seems clear that there has at no time been a connection between either of the meres at Hornsea and the sea.

A few years ago on the site of the gasworks, which are on the edge of the Mere, an excellent opportunity was afforded of examining the old deposits formed by the present Mere, and these again clearly indicated that the salt water had not at any time reached the lake.

Poulson informs us that formerly the waters from Catfoss emptied themselves into the Mere, but were diverted by the Rev. Mr. Constable.

Photo by]   [*J. W. Stather.*

THE SITE OF THE OLD MERE, HORNSEA.

Photo by]   [*J. W. Stather.*

PEAT CONTAINING FRESH-WATER REMAINS, BEING PART OF THE BED
OF THE OLD MERE, HORNSEA, 1906.

CHAPTER XXI

Atwick: its Cross—Cleeton—Skipsea: its Ancient Mere—Pre-Historic Earthworks—Hude or Huth—Ulrome—Barmston—Hartburn.

During 107 years the loss was two yards a year at Atwick (wrote Pickwell in 1878). The cross was measured from the cliff edge in 1786, by Tuke, and found to be about 980 yards * ; in 1840 Poulson stated it was " scarcely half that distance." In 1832 it was 885 yards, or a loss of 95 yards in 66 years. Mr. Morfitt tells me that in February 1912, the distance was 705 yards.

According to Mr. Matthews, the loss of the cliffs at Atwick in recent years has been 2½ to 3 yards per annum.

At Skirklington, between Skipsea and Atwick, the loss was 1·9 yards during 107 years (Pickwell in 1878). More recently, 2 yards per annum (Matthews).

Cleton, Cleeton or Clayton, according to some authorities, took its name from the nature of the soil—the " clay-town." It has now gone, but was once to the south-east of Skipsea.

" Cleeton lands " is a name given to different parts of Skipsea, and the name appears to be almost the only remaining record of this one-time township.

* Poulson (Vol. I., p. 174) says :—" 33 chains 61 links ' was the distance in 1786, but as the ordnance map of 1896 shewed the distance to be over 660 yards (30 chains) there must be some mistake here.

There is little doubt that formerly it was of more importance than Skipsea, and it seems fairly clear that in the Domesday Survey Skipsea was included under Cleton, viz., " Cleton or Clayton. In Cletune Harold had twenty-eight carucates of land, and one oxgang and a half to be taxed, where they may be twenty-eight

THE OLD CROSS, ATWICK.

ploughs. Drogo has now there two ploughs, and six villaines with one plough, and one hundred acres of meadow. To this manor belongs the soke in Dring-olme and Upton, five carucates of land and a half to be taxed, where there may be five ploughs and a half. There is now there one villaine having two oxen. The whole manor, with its adjacent parts, five and a half miles long, and one mile broad, value in King Edward's time thirty-two pounds, now six pounds."

Probably the lost township of Hythe was in the " adjacent parts " of this once five and a half miles parish.

A copyhold rental, dated 1768, shows that Cleton was then worth £32 to its various owners.

During 111 years the average rate of erosion of the land at Skipsea was (according to Pickwell in 1878) just under 2 yards a year.

Mr. Hatfield measured the distance from the wind-mill at Skipsea to the sea in 1833, and it was 1757 yards. This wind-mill has now gone, but I am informed by the Rev. R. W. Watson that its site was, in February, 1912, 225 yards nearer the sea than in 1833, which is equivalent to a loss of 3 yards a year for the 79 years.

In recent years the loss seems to have been at the rate of two yards per annum (Matthews).

The lacustrine deposit, or old lake bed,* which is now fairly well washed away by the sea, but which twenty years ago was seen in magnificent section in the cliffs, is certainly positive evidence of a mere or lake having existed in Skipsea. In this way the village once re-sembled Hornsea, Withernsea, Owthorne and other places. From an Inquisition held at Waghen about 1288, it seems that Rob. de Chester enjoyed the tythe of fish in Skipsea Marr. Possibly, therefore, this particular mere was in existence, and well stocked with fish at that

* See " Geological Rambles in East Yorkshire," p. 251.

time. It would certainly be some distance inland from the cliff edge in the 13th century.

Poulson has the following sentence in reference to Skipsea,* which is not quite as clear as it might be :— " Mr. Penant, the tourist; states that in his time large masses of amber are found here upon this coast, but it has disappeared about seven years ago."

PEAT AT SKIPSEA.

Close to Skipsea is Skipsea Brough, which contains a remarkable series of British earthworks, and a central mound, which were occupied by Drogo in the time of William the Norman, and upon which are yet the remains of his ' keep.' And in the adjacent village of Ulrome, within the margin of an ancient mere, now dry, Mr. T. Boynton,

* Vol. I., p. 45-6.

about thirty years ago, discovered some remarkable pile-dwellings, which yielded many important relics of the Stone and Bronze Ages.

A bone object, in shape resembling a spear-head, said to be of British date, " was found in the cliffs opposite to Skipsea Brough, embedded about six feet below the surface of the earth." It is figured by Poulson.

Hyde, Hide, or Hyth, " in Saxon a port or haven, expressive of its situation," was formerly to the east of

Photo by]   [*C. W. Mason.*

THE BRITISH EARTHWORK AT SKIPSEA.

Skipsea. Its precise position is unknown, except that its site is far out to sea.

The place is not referred to in Domesday, and, like Skipsea, was probably included in the five and a half miles of Cleeton. In the reign of Edward II. it is referred to, in association with Skipsea and Cleton, and even so early as in the reign of Edward III., Hythe was included in a petition to the king for a reduction in the assessments, in consequence of the devastations of the sea.*

In the year 1400, at an Inquisition held at Hedon, it appears that the convent of Meaux had been in receipt of tithes to the value of £46 13s. 4d. from Ulram, Cleton and Skipsea, which last included Villam de Hyth. And it is important to notice that of this amount (large indeed for that early time), no less a sum than £30 was received from Hyth, " chiefly in the tythe of fish . . . . . all destroyed."

The chronicler of Meaux, naturally bitterly complains of this serious loss, and as he informs us that in 1396 " the place is totally destroyed," it is not to be expected that there is much now left relating to the place. Its site is probably further out to sea than any of which we have a definite record.

Pickwell, in 1878, recorded that the loss at Ulrome averaged 1·2 yards a year. Mr. E. R. Matthews estimates that during the past few years the loss of land has varied from 1½ to 4 yards per annum.

Withow was originally included in the township of Cleton, and, beyond its name, we have little left relating to it. In the Inquisition held at Waghen in 1288, already referred to, Robert de Chester held the tithe of fish in Skipsea Marr, and in Withow. There are the remains

* Among the papers at Burton Agnes, recently described at a meeting of the East Riding Antiquarian Society by the Rev. C. V. Collier, reference was made to a *New Hythe* in the XVII. Century. This would seem to indicate that at that period there was still evidence of the hamlet that had taken the place of the *Old Hythe*, which had been washed away. "Skipsea cum Neuhithe" is also referred to in *Comitissa Cornubiæ*.

of an old lake on the coast, at one part of which remains of an Irish Elk, etc., have been found, and this hollow still goes by the name of Withow Hole.

At the south end of Barmston parish, there was a loss of 1·1 yards per annum between 1756 and 1876, or 132 yards in 120 years, whereas near Barmston outfall the loss was 126 yards in 120 years. In 1786 Tuke measured

Photo by]   [J. Hollingworth.

BARMSTON OUTFALL, LOOKING TOWARDS THE SEA.

the distance from the Townend Gate to the cliff as 790 yards, whereas in 1832 Mr. Hatfield measured the distance to be 745½ yards ; a loss of 44½ yards in 46 years.

Hartburn, or Hertburn. On the land joining the lost village of Hartburn, just south of Bridlington, Pickwell quotes a loss of 80 yards in 120 years, or two feet a year. With Poulson, we can only regret that practically no data exist by which the period of the destruction of this hamlet can be ascertained. On Tuke's map, 1786, it is merely recorded as washed away by the sea, whereas Dade calls it " a little vill. or tything, in conjunction with Winkton, depopulated and totally extinguished."

CHAPTER XXII

AUBURN—AUBURN HOUSE—WILSTHORPE—EROSION—
BRIDLINGTON : CHANGES NORTH AND SOUTH OF THE
HARBOUR—VARIOUS PIERS.

FOR many years the township of Auburn was constantly
assailed by the sea, its site being now some distance from
the cliff edge, and covered by about 16 feet of water

Photo by]   *[J. Hollingworth.*
ALL THAT IS LEFT OF THE VILLAGE OF AUBURN.

at high-water spring tides. When the last building of
the village was reached, and part of it was washed away,
the sea seemed to stay its work, sand-dunes were formed,
and though I have been familiar with "Auburn House"
for nearly a quarter of a century, it does not change at

all, and the part remaining has been bricked up and inhabited.

A few months ago, a slight change in the position of one of the sand-dunes a little north of Auburn House,

IRON MILESTONE, ONCE ON THE ROAD FROM BRIDLINGTON
TO BEVERLEY, NOW WASHED AWAY.

revealed a fragment of the old coaching road to Beverley, most of which is now washed away. I obtained a number of horse-shoes, etc., and, buried among the sand, was the old iron "milestone" (the post which formerly

197

supported it had long since decayed) informing the traveller the distance to Beverley. This is now at Hull, and the accompanying illustration is taken from a photograph of it.

From the farmhouse, Wilsthorpe, near Bridlington, to the cliff edge in 1810 was 221 yards. In 1833 Mr. Coverly re-measured, and found it to be about 180 yards—a loss of 41 yards in 23 years.

In 1905 Mr. Matthews reported that all the village had disappeared, with the exception of one house, which was then six feet from the cliff edge, though a few years previously it was some distance from the cliff. A few waterworn bricks on the beach are all that now are left of this house.

The name Bridlington is now given to what was formerly the Quay or "Key" of Bridlington or Burlington. The "old town" of Bridlington, as it is now called (and pronounced locally as "Bollinton" or "Bolliton") is a mile inland. In recent years the place has developed enormously as a pleasure resort, and as a residence for Hull and Leeds merchants.

"The Quay" doubtless originated as a result of the natural harbour or creek formed by the sea-ward extremity of the Gypsey Race, which still enters the harbour there.

At one time the erosion of the cliffs immediately to the north and south of the pier was very severe, but in recent years the Prince's Parade, Promenade, Spa, and other structures with strong protective sea-walls, as well as

[*Spurr.*

BRIDLINGTON IN 1868, ON THE SITE OF THE PRESENT PROMENADE.

The building (Fort Hall), marked by an arrow, is similarly shown on page 201.

the groynes and other means of defence, have practically
stayed the work of the sea along the front of this borough.
Unlike the other places on this coast, the greatest changes
at Bridlington are, that the sea-front has really extended
seaward by these artificial additions. These changes are
perhaps best illustrated by the accompanying reproductions of old engravings and photographs.

BRIDLINGTON QUAY IN 1830.

The foundations of the small jetty on the right can still be seen
at low water.

(*From Allen's " Yorkshire "*)

Immediately south of the Harbour at Bridlington,
Mr. Pickwell* gives the loss between 1805 and 1852
as two yards per annum ; between 1852 and 1872, $3\frac{1}{2}$
yards per annum, whereas between 1872 and 1885 the

* " Proc. Inst. Civ. Eng.", 1878, pp. 191-212.

Photo by]

Bridlington in 1910.   (See also page 199).

[Spurr.

loss, as estimated by Mr. Lamplugh, has been 5 yards a
year. The erection of groynes north of the town has
resulted in the land to the south suffering. Between
1805 and 1885 therefore, the total loss was 229 yards.
At Hilderthorpe, which is immediately adjoining (being

BRIDLINGTON HARBOUR, 1837.
(From Water-colour by Copley Fielding in the Wallace Collection).

part of Bridlington), Mr. Pickwell found the loss between
1805 and 1852 to be 2·1 yards a year.

North of Bridlington, between the promenade and
Sewerby, Mr. E. R. Matthews estimates the erosion at
six feet a year.

There is an interesting plan of Bridlington in Knox's

BRIDLINGTON IN 1846, FROM THE SOUTH.

BRIDLINGTON IN 1846, FROM THE NORTH.

"Eastern Yorkshire," which shows the positions of the two "old piers" and of the present "new piers," the date of which is presumably indicated by the words "Enlarged Harbour, 1847." The plan shows the position of the "Fountain," and also the "Hull Road," which,

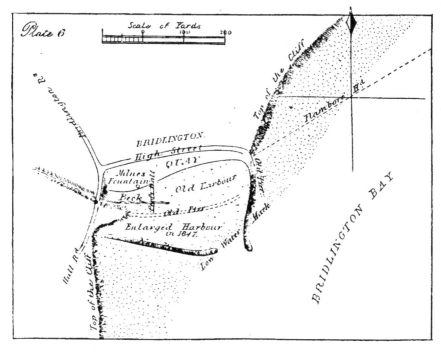

PLAN OF BRIDLINGTON HARBOUR IN 1855.

Showing the position of the "Old Pier" prior to 1847, and also showing the "Hull Road," now washed away.

(From Knox's "Descriptions Geological, etc., of Eastern Yorkshire").

a little to the south of this point, is now washed away by the sea.

On the same plate is a plan of Filey and Filey Bay, with "Suggested Piers for a larger or a smaller Harbour by R. Knox," which, needless to say, were not carried out.

BRIDLINGTON FROM THE SOUTH, *circa* 1850.

BRIDLINGTON FROM BESSINGBY HILL IN 1854.

BRIDLINGTON, SHOWING THE OLD PIER.

(From an old print).

CHAPTER XXIII

Lord Burleigh's Chart, *Temp.* Henry VIII.—
Former Configuration of Holderness—Beacons
—Armada Scare—Methods of Lighting the
Beacons.

In the British Museum is an admirable manuscript
chart of East Yorkshire, which was formerly in the
possession of Lord Burleigh. It is inscribed " A Plotte
made for the discripcion of the River of Humber and of the
Sea and Seacoost from Hull to Skarburgh in the Reigne of
Queene Elizabeth," and was obviously prepared in connection with the Spanish Armada scare. Some fac-
similes were issued by Messrs. Peck & Sons of Hull.

The original is a gorgeous parchment with decorated
border in red, blue, yellow, and gold ; in the top right
hand corner is the following description :—

" This plotte ys made for the discripcion of the river
of Humber and of the sea and seacoost from Hull to
Skarburgh, wherfore though sum [some] hamlettes and
villages of Holderness be left oute. It is not materiall
for the purpose this was made.

" Note that the numbres and figures next the shippes
signifye the depthe of the Channell at lowe water accompt-
ing by fathomes.

" Humber Banke on Holdernes syde from the South
blockehous at Hull to Esington eastward ys commonly
one yarde and a haulfe or ij yardes of height, saving at

Paule Hill, and there the banke is higher alongist the Shore by haulfe a quarter of a myle towardes Humber mouth.

" All alongist the shore of Holdernes, saving three quarter of a myle foranempst [afore-named] Pawle ys Clay grounde or woes [ooze] and that ys pebill stones.

" All the wyndes (winds) be noted where the shippes do ryde to declare what roods be best in anny wynde.

The Depth of Hull Water at the bridge nere ye fortes.

At an nepe ....viij foote ....Full sea at nepe iij fathom.

Att spring .... vj foote ....Att Spryng....iij fathom.

Betwixt theast and weste neste.

At an nepe at lowe water..iiij foote..full Sea at nepe iij fathom.

Att Spryng........iij foote..At Spryng....iiij fathom.

Breaches on Humber Bancke on Holdernes side.

" Two breaches at Sawtey [Saltagh] and one at Welwicke called Bitchecrofte as large as both thother ij ffor the repayring whereof the Quenes Maiesties Landes be chargeable, And therfore the said breaches ought to be made and amended at hyr Highnes Coste expences and charges, or ellse the countrey and hir Maiesties Landes there wilbe drowned."

In the water are various types of sixteenth century vessels, varying from examples of the largest types to rowing boats, all drawn quite regardless of perspective. On the land the various churches, monastic buildings, parks, and woods are indicated in colours. Commencing at the north extremity is " Skarburgh " and its castle, with a beacon on the south. Just north of Flamborough headland a further beacon surmounts a hill. At " Righton " (Reighton), at Bridlington (where two piers are shown) there are three beacons, and between Barmston and

LORD BURLEIGH'S CHART.
(Temp. *Henry VIII.* *British Museum*).

Skipsea there are three others ; south of Hornsea there are three ; there are three opposite " Mapilton " and there are three between " Awbrough " [Aldborough] and " Grymston Garth " ; a further beacon is shown at Withorntsea [Withernsea] ; there appear to be no others on the coast until we get into the Humber at Pauleholme, where three beacons are shown. Oddly enough, nothing of the kind is shown at Spurn Point. Along the coast-line there are many places represented which have almost gone, and some which have since entirely disappeared. On the Humber shore, Burstall Priory is clearly indicated, and in front of it the water is sufficiently deep for the anchorage of the fairly large ship which is represented. Within the Spurn peninsula three ships are anchored, and just north of Withernsea is a well-defined bay, and another south of Tunstall (Sand-le-Mere) with the following description :—" These small ereke for landing of fysher boote wherein small shyps at spryng tydes may also enter and do annoyaunse." Between these two bays Owthorne Church with its square tower is clearly represented. "Horneseymar" is shown at some distance inland, and it is connected with the sea by a wide stream, to the north of which a fairly large pier is represented. Just north of Earlsdyke is Auburn, with what appears to be a double beacon, though it is to the west of the village. Willsthorpe is also shown, and Sowerby.

With regard to the beacons shown on Lord Burleigh's chart, a letter dated June 21st, 1558, was sent to the justices of each riding in Yorkshire by Queen Elizabeth, in consequence of which a certificate of the state and number of the beacons in the district was made, from which the following extracts, bearing upon the coast, are made :—

" Dickering, XIX Beacons.

Bridlington cum Key, three beacons uppon the sea cost, geving lighte to Flambroughe and Fraistroppe.

Flambrough, three beacons uppon the sea cost, takinge lighte from Bridlington, and geving lighte to Rudstone.

Muston, three beacons, half a myll from the sea cost, taking lighte from Righton, and geveth lighte to Staxton.

Righton cum Spetonn, three beacons on the sea cost, takinge lighte from Flambroughe, and geveth light to Rudstone

Rudstone, two beacons foure mylls from the sea cost, taketh lighte from Flambrough and Righton, and geveth lighte to Ruston.

Ruston beacon, six mylls from the sea cost, taketh lighte from Rudstone, and geveth lighte to Bainton beacon.

Fraistroppe cum Awburne, three beacons, a myll from the sea, taketh lighte from Bridlington, and geveth lighte to Houlderness.

Stanton beacon, taketh light from Muston, and geveth light to Coleham.

Buckrose II. Beacons.

At Coleham, one beacon, takinge lighte from Stanton and Bridlington, and geveth lighte to Settrington. Some affirm that yt may be sene as farre as Hornsey in Houlderness.

Settrington beacon taketh lighte at Coleham and Scarbrough, and geveth lighte to Whitwell beacon, and all that way to Yorke, and into Harthill, and over the most part of Pickeringe Lythe.

Harthill, VI Beacons.

Hunsley, two beacons, takinge lighte from Bainton, and geveth lighte to Holme.

Bainton, two beacons, takinge lighte from Ruston, and giveth light to Hunsley and Wilton.

Wilton beacon taketh lighte from Bainton, Hunsley, and Ruston, and giveth lighte to Holme beacon, to the cytty of Yorke, and to the lowe Cuntreye.

Holme beacon taketh lighte from Hunsley and Wilton, and geveth lighte to Marchland and the Lowe Cuntreyes.

Houlderness.

Kilneseye, three beacons
Dimilton, three beacons
Withernese, three beacons
Waxholme, three beacons
Grimeston, three beacons
All uppon the sea cost, and do geve lighte to these beacons followinge uppon Humber.

Welwicke
Patrington
Bowerhouse-hill
Paulle
Marfleet
These beacons standinge uppon Humber, and takinge lighte from the beacons aforesaid, do geve lighte to the beacons in Hartil and the Hulshier, which are Transbye and Hunsay.

Further upon the sea cost in Houlderness.

Awbrough, three beacons
Mappleton, three beacons
Hornsey, three beacons
Skipsey, three beacons
Barnstone, three beacons
Those give lighte to Bainton and Rudstone.

Owse and Darwine [Ouse and Derwent].

In Ouse and Darwine, it is certified that there are not anie beacons.

*Orders to be observed for the watching of the
Beacons in Houlderness and other parts of the East-Ryding*

To Fyre one Beacon.

First, that the beacons be watched dilligentlie, and
that there be a certain number of the wisest and discretest
men dwellinge within the limits of everie beacon place,
one of them at the least every daie, and every night, being
assotiate with a reasonable number of other persons ;
that is to say, that there be everie daie two persons, and
everic night three persons, and none be allowed to watch
but honest householders, and above the age of 30 yeres,
except he be specialli elected and appointed by name.

Item ; that yff the said watchmen se or discerne anie
shipps on the sea, or in the ryver of Humber, which by
there stay, or alteration of there course or otherwaise,
they geve plain occasion of suspicion to be enemies, and
doubtful to do some harme eyther on the maine lande, or
to some of oure shippes saylinge alongst the coste on the
sea ; that then the said watchmen, with the advice and
direction of some of the wisest and discretest men appointed as is aforesaid, shall sett on fire one of the beacons
onlie, to the intente not onlie to give oure shippes on the
sea warninge of the enemie, but also to be a warninge to
the people of the cuntrey, that they loke the better about
them, and be the more redie as occasion shall serve, and
that none of the beacons within the lande (where there
is but two together) shal be sett on fyre, for the fyreinge
of one beacon onlie uppon the shore.

To Fyre two Beacons.

Item ; if the said watchemen do see or discerne anie
great number of shippes, which by the course aforesaid,
or otherwise, do geve vehement suspicion to be enemies,
and to be doubted that they meene to invade ; that then,
they shall with the advice aforesaid, sett two of the said

thre beacons (standing together) on fyre, where uppon the watchmen with in the mayne land, where two beacons be together, shall onlie sett one of the said two beacons to the intent, that every man in charge put himself in armoure and be redy.

To Fyre three Beacons.

Item; yf the said watchmen on the sea coste (or Humber) do see any greate number of shippes which to them is apparant to be enimies, and that they offer and do come of land to invade, that then the said watchmen (where three beacons be together) shall sett all there three beacons on fier, upon sight whereof, the watchmen within the land (where two beacons be together) shall fyer bothe there beacons. And, that farther within the land (where there is but one beacon in the place) the watchmen shall sett the same on fyer, to the intente to geve warninge and knowledge of the present danger, and that there uppon, every man in charge may resorte with all speed to the place or shore from whence the first lighte was geven, as shall be thought meet by the General or chiefe leader.

Item; if there be anie occasion of fieringe of beacones within the east end of Houlderness or alongst the sea cost that then every of them take there light at others as followeth; from Kilnsea to Wellwike beacones; from Wellwike to Patrington beacons; from Pattrington to the beacone field or the Bowerhousehill beacones; from thence to Pawll to Marflet beacones; from Marflet (which is the westermost beacone in Houlderness) the two beacones of Tranlie shall take there lighte from Paull, or other the beacones on the shore of Humber and geve light to the Holme beacone; and from Holme beacone west warde.

Item; yf anie accasion, as is aforesaid, of fyeringe of beacons be on the Northe parte of Houlderness, or within the weapontake of Dickering, where there beacons stand together, that then they shall take there lighte one at another, viz. : from anie of the beacons on the sea cost or shore to Staxton beacons, from Staxton to Settrington beacons, from thence to Whitwell beacon, and also from Hornsey and Barmeston, or any of the beacons on that parte.

Rudstone beacon shall take his light from Rudstone to Coleham to Wilton, and so into the Lowe Cuntrey ; and likewise from Hornsey or other the beacons on the shore. Baynton beacon doth take his lighte from Baynton to Wilton, and so towards Yorke and the upper parte of the countrey.

Item ; that there be provided for everie beacon half a tar barrell at the least ; and for everie three beacon standing together there, three half barrells to spare, tc the end that yf by occasion, anie of the barrels be spent by reason of fyeringe of one beacon onlie, or two (upon that small occasion) that then the reast may be redy for that purpose, when they shall be forced to sett more than one on fyer ; and that the same barells be left in the kepeinge of some honest, substantial man, dwellinge niest or neare unto the said beacons.

Item ; that the people of the contrey have warninge that the same invasion be on the sea coste, that then all the catell, shepe, horse, and victual, be carried, and driven from thence where the enimie shall be, into the mayn land, and that everie man do his best endevors to the uttermoste of his power to prevent the enemie of all victualls, or other commodities which by anie meanes maie stand them in steade."

CHAPTER XXIV

Maps of East Yorkshire, *temp*. Henry VIII—Saxton (1577)—Speed (1610)—Bleau (1662)—Blome—Ogilby (1675)—Collins (1684)—Morden (1722)—Moll (1724)—1725—Scott (1740)—Bowen (1750)—Jeffreys (1772)—Kitchen (1777)—Fayden (1780)—Tuke (1786)—Cary (1805)—Trinity House (1820)—Greenwood (1834, 1838, 1840)—Hall (1846)—Boyle (1880).

In addition to Lord Burleigh's chart just described, the following selection has been made from a collection of scores of maps and plans of the district referred to in these notes. To give a bare list would occupy too much space.* A few, therefore, representing different periods, are given, and will illustrate the changes that have taken place since the sixteenth century.

MS. Map, *temp.* Henry VIII.

This interesting chart is in the British Museum, and is of the time of Henry VIII., though after 1541, as the castle and blockhouses on the east side of the river, shown on the chart, were only erected in that year. Allowing for the exaggeration of glacial mounds and rounded low chalk wolds into miniature Matterhorns, the membrane

---

* See "List of Papers, Maps, etc., relating to the erosion of the Holderness coast, and to changes in the Humber estuary," by the present writer. Trans. Hull Geol. Soc., 1906.

PLAN OF THE HUMBER, *Temp*. HENRY VIII., SHOWING AN ISLAND OUTSIDE SPURN POINT.

has one or two points of particular interest. Chief among these is the fact that an island is distinctly represented to the east of Spurn Point, and there is another in the centre of the Humber between Spurn and Clee-thorpes. The view of Hull with its two bridges across the river is also of interest, and it will be noticed that Patrington is connected with the river by means of a channel, which has long been silted up.

SAXTON, 1577.

The earliest engraved map of the county is apparently by Christopher Saxton, a Yorkshireman. Several editions of his maps were published, though the first appears to be dated 1577, and is consequently much earlier than Speed's better-known maps, though these appear to be largely Saxton's, " brought up to date." Saxton died in 1587. His maps were re-printed from the original plates in 1645, and were " revised " by omitting the arms of the gentleman at whose " charge " the maps were originally " graven for the public good," and by altering the date ! By the courtesy of the officials at the British Museum, I am able to give a reproduction of the first edition of Saxton's map—the only copy I have seen.

JOHN SPEED, 1610.

John Speed's Map of " The North and East Ridings of Yorkshire," appearing in his " Theatre of Great Britain," is one of the best we have of the district at this

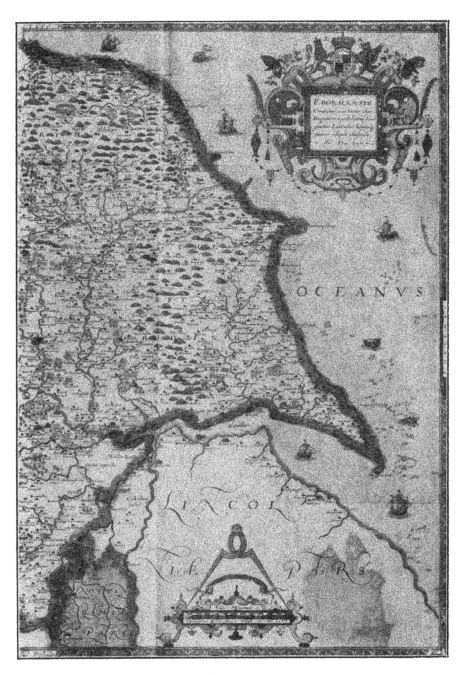

PORTION OF SAXTON'S MAP, 1577.

The earliest Engraved Map of Yorkshire.

(*From the copy in the British Museum*).

early period.   It shows no trace of Sunk Island.   Spurn

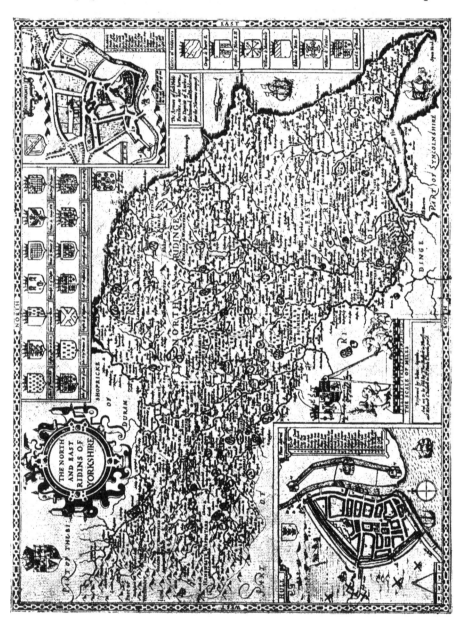

SPEED'S MAP OF THE NORTH AND EAST RIDINGS, 1610.

has the characteristic broad and short outline indicated in
maps of this period and later.   A broad entrance con-

neets Hornsea Mere with the sea, and on the west it is united with the River Hull by one of its tributaries. A stream is also shown joining a mere at Waxholme with the Humber. A valuable plan of Hull at this time appears as an inset to this map, of which there are two or three

BLAEU'S MAP, 1662.

(In "Geographiæ Blavianæ volumen quintum quo Anglia . . . continetur," Amsterdam).

editions. There is also one of the whole county by the same engraver.

BLAEU, 1662.

This map is often gaudily coloured, and shows Spurn Headland rather short, and no trace of Sunk Island, nor of any other island whatever in the Humber. From

Hornsea Mere there is a somewhat lengthy entrance to the sea, which is probably a representation of the one-time important harbour at that place. As already pointed out, however, it seems hardly likely, judging from the recent and semi-fossil fauna and flora, that Hornsea Mere was ever actually connected with the North Sea. The river rising in a lake at Waxholme is interesting. The present site of the mere at this point is, of course, in the sea, though traces of it may still be seen on the beach.

RICHARD BLOME, 17TH CENTURY.

Blome's somewhat crude map dates from the early part of the seventeenth century. Hornsea Mere is represented as a lagoon with a very wide entrance from the sea, and is also connected by a stream with the River Hull. Spurn has the characteristic short and broad appearance usually shown on these early maps.

OGILBY (ROAD MAP) 1675.

In the seventeenth century, and later, several road-maps were published for the aid of travellers. These appeared on scrolls, after the manner of the modern motor maps. The miles were indicated on the roads, and in addition, beacons, gibbets, brooks, bridges, churches, mills, hills, moors and commons were shown. The section given herewith (page 223) shows part of the plan of " the road from London to Flamborough," viz., between Kilham and the sea. These maps were engraved

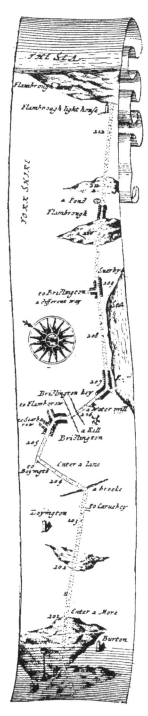

by Hollar between 1675 and 1698. From the title of one of the maps we have reproduced the illustration

FROM THE ENGRAVED TITLE OF OGILBY'S ROAD MAP, SHEWING THE PEDOMETER, BY MEANS OF WHICH THE DISTANCES WERE ASCERTAINED.

of the pedometer, etc., by means of which the road maps were made.

GREENVILE COLLINS, 1684.

This is a navigator's chart, but still has many points of interest. It

PART OF OGILBY'S ROAD MAP, SHEWING THE OLD LIGHTHOUSE AT FLAMBOROUGH, AND THE BEACON AT BURTON AGNES.

will be noticed that the " Den " west of Spurn is
definitely an island, and not merely a sand-dune or

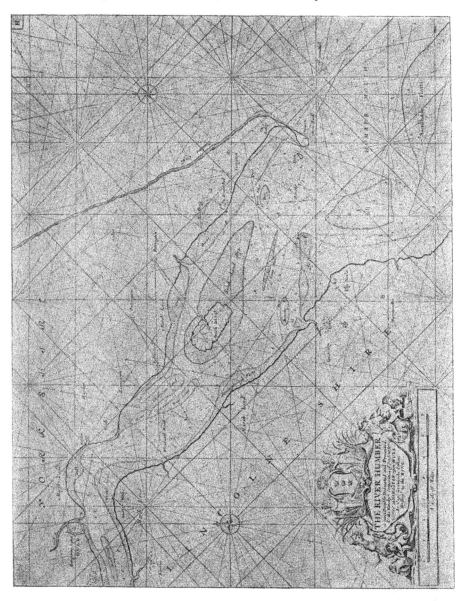

GREENVILE COLLINS' CHART, 1684.

mud-bank. The representation of Sunk Island also
shows the original enclosure.

ROBERT MORDEN, 1722.

Morden's map of 1722 shows Spurn as short and broad without the narrow isthmus and spoon-shaped extremity which appear later. Sunk Island is distinctly an island, with no connection whatever with the mainland. Hornsea Mere is joined to the sea by

MOLL'S MAP, 1724.

a broad waterway, and forms a lagoon, while on the west it is joined to the River Hull by a stream, over which, just west of the mere, is a bridge.

H. MOLL, 1724.

This is even more crude than the preceding map, but similar in detail. Spurn presents the shorter and

broader form already noted, and Sunk Island has no connection with the mainland.

1725.

A chart of the Humber appearing on the MS. Plan
of Hull, dated 1725, in the British Museum, in addition

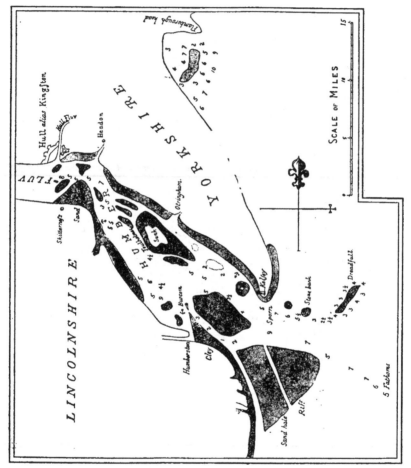

CHART SHEWING THE POSITIONS OF SUNK ISLAND AND THE
HUMBER SAND-BANKS IN 1725.

to showing Sunk Island still separated from the mainland,
also gives a small island west of Spurn, which place,
strange to say, is only indicated by the word " Sporn "
to the south of " Kelsey," where a lighthouse is shown.

LOST TOWNS OF THE YORKSHIRE COAST

JOHN SCOTT, 1734-1740.

Near the centre of this Chart, in a curiously decorated border, supported by two fish-tailed Neptunes, in the crowns of which water is spurted from the nostrils of two cross-tailed dolphins, is inscribed " This Draught of the RIVER HUMBER is most humbly Dedicated to the

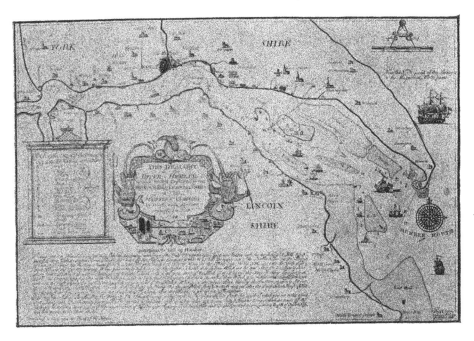

SCOTT'S MAP, 1740.

Various editions of this map were issued, the chief difference of which was the date, which was often filled in with ink.

Honourable Commissioners of his Majesty's Customs by Their most Obedient seruant John Scott, 1734."

It is a chart of unusual interest. On the short and north-westwardly curved Spurn are the two " lighthouses." The larger, apparently a square brick building, has a long pole extending from the top, at the end of which

227

is an iron cage-like beacon which held the fire ; the smaller and southerly light was evidently of wood, with an arm, gallows-like, at the end of which was the beacon. The half-moon shape of the Spurn as represented is different from that shown on any other chart (see page 69). A little to the north-west is a circular patch labelled " Old Spurn."

A small island labelled " Sunk " is joined to the mainland near " Headon " by a mudbank labelled " Holmes." It is shown as about 3 miles long and about $1\frac{1}{2}$ miles wide, and there are half-a-dozen buildings upon it. A shallow channel extends to the right of Sunk, towards " Potterfleet," south of " Otteringham." Skitter Ness, on the Lincolnshire coast opposite Marfleet, diverts the channel northwards. " K. S. Hull," with its church and two small jetties and a dolphin, is clearly shown, with the Charter House and Sugar House on the north, and the three blockhouses and Drypool Church on the east. At Hessle a triangular mudbank diverts the channel to the south again, and nine other smaller mudbanks are in the middle of the estuary between Ferriby and Blacktoft, and a small one joins the Yorkshire shore at Oyster Ness.

At the bottom of the Chart are the following curious " Directions to sail up Humber " :—" Having got into Humber in a fair way between the Spurn, and Buoy of the Bull (keeping the lesser Light about an Handspoke length open to the Southward of the higher Light to avoid Trinity Sand), the Course by the Magnetic Compass is W. N. W. $\frac{1}{2}$ Nly. to the upper end of the Burkham leaving the Buoys of Clea Ness, and the Burkham upon the Larboard Side you must take care to allow for the Flood Tide Setting N.W. and near the North Side more Nly. and the Ebb the contrary. When you come so high up as Kayingham Church, and Salton Wood are in one, Steer for the Lincoln Shire Shore (The Channel laying pretty nigh it) till you come so high that Ottringham Church is on with Salton Wood you will perceiue Paul Town open open [sic] with its Jetty : and Marfleet Church with its Jetty on : (being the thwart Mark for Skitter Ness end) as also the South Block House of Hull and the Jayl on : you are in a fair way between Skitter Ness Sand and the Hebles, to go up into Hull Road. In Sailing above Hull, Steer for Barrow West jetty (the Channel laying pretty nigh it) to avoid Hessle Sand ; the point of which you are rather above when Cottingham Church, and Beverley Minster, as also a few Trees nigh the River Side, are in one (which Minster being without the limits of the Draught is not laid down). Then following the Tract of the Draught ; having Swanland Mill and an Hedge End in one you are Length of Oyster Ness and then go on as the Tract directs."

Added to this, in a contemporary handwriting, evidently within a year or two of its publication, is the following :—

" Since this Draught was publisht, the lower Light at the Spurn was removed December 1735. So that it now beares with the high Light overend N. W. by N. (therefore not now a Mark for Trinity Sand). The presant place of the low Light is thus markt ⊡. A Buoy is also laid upon the Hook of the Holmes."

The Chart is " Engraved and Printed by J. Hilbert, Hull." The scale is a trifle over half-an-inch to a mile, and the engraved surface 20 inches by 14 inches.

BOWEN, 1750.

Bowen's map of 1750 is very carefully. drawn, and
confirms the evidence supplied by other maps of the

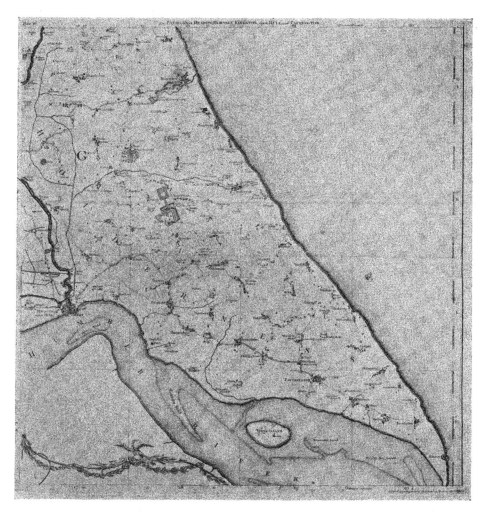

JEFFREYS' MAP, 1772.

period that Spurn was shorter and broader than it is
to-day. A " Sunk Sand " is shown to the east of Sunk
Island, the position of which is about the centre of the

river, broad navigable channels apparently existing on each side. The view of Hull from the Humber is one of the most valuable pieces of evidence we possess concerning the appearance of the town's frontage at that date. The map is reproduced on page 13.

THOMAS JEFFREYS, 1772.

This somewhat scarce map, unfortunately, does not show the tip of Spurn Point, but it illustrates very well the small size and position of Sunk Island, and is also instructive from the names of the number of places which existed on the coast-line but have since disappeared. In the same atlas is a small key map from which it is possible to fill in the outline of Spurn.

THOMAS KITCHIN, 1777.

This map also shows Sunk Island as an island, and around it " Sunk Sand," which later was attached to the mainland.

FAYDEN, 1780.

This map, which appears in Smeaton's Report, is figured and described on pages 79-80.

J. TUKE, 1786.

One of the most useful plans for our purpose is that issued by J. Tuke in the year 1786. This gives the actual distance from the sea, at that time, of several villages near the coast. These are as follows :—

Barmston, from the gate at the east end of the village   ..   ..   792 yds.

			yds.
Hornsea from the Church	..	..	1,133
Aldbrough ,, ,, ,, ..	..	..	2,044
Tunstall ,, ,, ,, ..	..	..	924
Holmpton ,, ,, ,, ..	..	..	1,200

The map is on a large scale, and, in addition to showing Pensthorpe, Birstall Garth and Priory, Owthorne, Sandley Meer, Auburne, and other places then existing, but which have since disappeared, it also shows " Hartburn, washed away by the sea," " Hyde, washed away by the sea," " Site of the town of Hornsea Beck," " Site of Hornsea Burton," " Site of the ancient Church of Aldbrough," and " Site of the ancient Church of Withernsea," on the sea-coast ; while within the Humber Estuary we find " Site of the Town of Ravenser," " Site of the Town of Ravenser Odd," and " Site of the Town of Frismarsh." This plan shows the development of Sunk Island, being joined to the mainland on the west, but with a great stretch of mud-bank to the east.

CARY, 1805.

This exhibits evidence of greater care having been taken in its execution, and can be relied upon as being fairly accurate.

TRINITY HOUSE CHART, 1820.

This is issued together with " Directions for Ships Sailing by the Spurn Floating Light," which had then been moored off the mouth of the Humber. The chart shows two distinct breaks in the neck of land forming Spurn Point, between Kilnsea and the lighthouses. To

TUKE'S MAP, 1786.

the south-east of the Spurn lights a sand-bank is marked as " dry," while two others, more towards the east, are " nearly dry." The chart is 8½ inches by 12½ inches, is engraved by D. Henwood, and was published " by order of the Trinity House, London, 1820 " (see page 235).

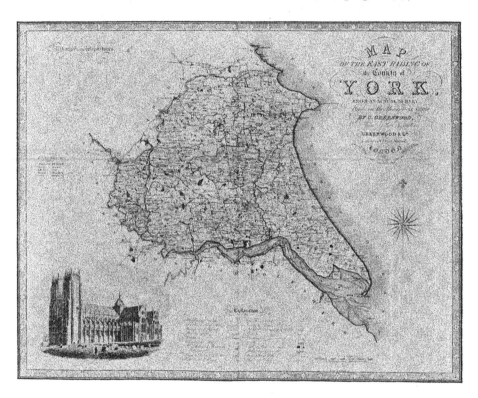

GREENWOOD, 1834.

GREENWOOD, 1834.

This fine map, from a careful survey, contains many important features in the Humber area. It also shows much of Sunk Island joined to the mainland.

1838.

A map of 1838 shows the present road leading from

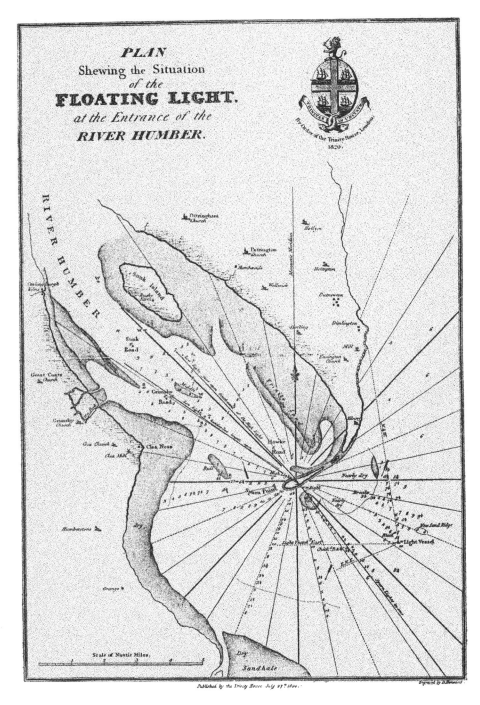

CHART SHOWING TWO BREAKS IN THE LAND AT SPURN, 1820.

Easington to the sea as continued along the cliff to Kiln-sea. The cliff road has now entirely disappeared, the present road to Kilnsea being inland. Hornsea Mere must have been unknown to the designer, as it is not even indicated.

1840. THE NORTH DIVISION OF HOLDERNESS.

Shows the coast between " Fraysthorpe " and Ald-borough. A beacon is shown on Hamilton Hills, just south of Earl's Dyke, and there is a " Bridle Road " near the cliff edge between Barmston Drain and Atwick, which has since disappeared. On the cliff edge at Ulrome is a " Coastguard Watch House," and there is a kiln at " Whity Hole " on the cliff at Skipsea. Moor Hill Beacon is shown on the cliff edge a mile north of Atwick. At Hornsea a number of lanes are shown, which have since been washed away ; and a mile north of Aldborough is Bunker's Hill Beacon. Engraved surface, 15×17. Scale 1 inch to a mile.

JOHN HALL, 1846.

A Chart of part of the Rivers Humber and Ouse, from Hull Roads to Goole. From a Survey taken by John Hall. (Second Edition).

Amongst the " observations " we notice " It is high water at Hull, full and change of the moon at 6 o'clock, and at Goole at half-past 8 o'clock, and the tide flows from 9 to 11 feet at Goole in the neap tides, and from 14 to 16 feet in the spring tides.

" The flood tides run up from $2\frac{1}{2}$ to 3 hours, both flood and ebb run with great rapidity.

" The channels are subject to alter and lay up in the summer months, and deepen by the winter freshes, which makes the navigation dangerous.

" Vessels bound to Goole, should be careful to wait until the tide is sufficiently flowed, so as they may be able to pass the Shallows.

" The port of Goole extends to the lower part of the River Ouse, the boundary is . . . from Foxfleit Ness to Bosom Cross."

Along the course of the river, which is represented from a little east of Hull, to Goole, various soundings are given, and good anchorage is also indicated. " Skitter Sand " is shown on the Lincolnshire shore east of Goxhill Ferry. To the east of Hessle a large tongue extends two-thirds away across the estuary, and reaches nearly to the mouth of the Hull ; there being a channel to the south only ; between North Ferriby and Ferriby Sluice (Lincolnshire), in the centre of the widest part of the estuary, is " Red Cliff Sand " and " Old Warp," with channels both on the north and south sides. " Pudding Pie Sand," between Oyster Ness and Brough, and " Whitton Sand," opposite Broomfleet, have also channels on each side of them.

The Chart was published by the author and engraved by Consitt and Goodwill, No. 15 Lowgate, Hull." Scale $\frac{3}{4}$ inch to a mile. Engraved surface 18 × 11 inches

LOST TOWNS OF THE YORKSHIRE COAST

BOYLE, 1889.

This was published by the late J. R. Boyle, and illustrates the positions of the lost towns of the Humber, which he ascertained as the result of his researches. It will be noticed that he has placed the positions of Ravenser and Ravenser Odd to the west of Spurn. The other places named are Sunthorpe, Penisthorpe, Tharlsthorpe, Frismersk, East Somerte, and Orwithfleet (see page 49).

CHAPTER XXV

ORIGIN OF THE HUMBER MUD—FROM THE RIVERS—FROM
THE SEA—EVIDENCE FROM THE MICROSCOPE—EFFECT
OF THE TIDES—ROYAL COMMISSION EVIDENCE.

THERE is no question that the waters of the Humber
are surcharged with particles of mud and sediment,
more so than in the case of any other English estuary.
The merest glance at the turbid stream will convince
anyone of the fact. While this sediment is in some cases
of value, and can be put to profit, at others it proves an
unmitigated nuisance, and a continual source of anxiety.
The Humber Conservancy and other bodies are kept
busily employed in watching and mapping the ever-
varying sand and mud banks. On the other hand, a
deposit of a similar nature, viz., the cranch at the mouth
of the Old Harbour, seems particularly obstinate in
occupying one position, and cannot be induced to follow
the example of its fellows amid stream. In the case of the
cranch, artificial means are employed to remove it.

When we come to the question of the source of the
sediment, however, there seems to be that diversity of
opinion which characterizes nearly all subjects of this
nature, and this difference of opinion is not an unhealthy
sign. Geology, in common with everything else, would
be dreary and dull indeed were it devoid of disputes and

consequent discussions, and it is only by proper discussions that the truth is, as a rule, arrived at.

The Humber mud can only have two sources :| it
must either come in from the various rivers draining

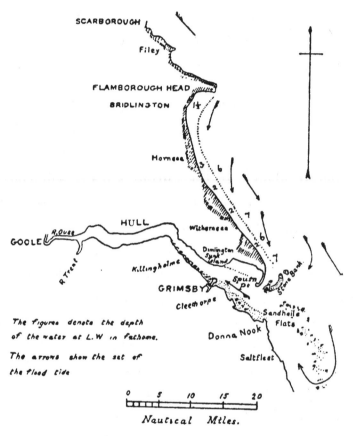

MAP SHOWING THE DIRECTION OF THE CURRENTS WHICH BRING
THE MUD DOWN THE COAST AND INTO THE HUMBER.

(After Wheeler).

into the estuary, or must reach the estuary through its
connection with the North Sea. Or, of course, the mud
may be derived from both the rivers and the sea. There
are persons who assert that it is entirely derived from

the rivers. Others are equally certain that it largely originates from the east coast. The champions of both these views not unnaturally believe that their respective theories are correct. Personally, I think that both sources contribute, and will here examine the facts.

It will be admitted that the mud in the Humber is accumulating. A most notable example is at Read's Island, between North and South Ferriby. Thirty years or so ago this was a comparatively small island, with a plot of grass in the centre on which a few cattle were reared. Now the island is hundreds of acres in extent, and has an enormous number of cattle grazing upon it. In the neighbourhood of Spurn and Sunk Island, Broomfleet Island, and in other parts of the Humber, acres upon acres of new land are continually being formed, while the Humber Channel itself is almost choked with sand and mud banks.

If the rivers Ouse, or Trent, or Hull, be examined at ebb tide, when the normal amount of water is flowing down, it will be seen that the waters are comparatively clear, and, though a fairly large amount of material may be brought down from the higher reaches of the rivers, *in solution*, very little appears to be coming down in suspension. It is the latter which mostly affects the question. It is a significant fact that when the tidal waters are flowing up stream, the water is much more muddy and coffee-coloured than when the waters of the rivers only are descending in their channels. Undoubtedly a quantity of detritus is carried along by the rivers, but this is mostly derived from the high ground near their sources, and the bulk of it in all probability goes towards building up the alluvial flats so characteristic of the Ouse and Trent.

These two rivers, but perhaps more particulaly the latter, are at times very muddy in their courses, especially when the water is flowing a little quicker than usual. The material which causes this is principally derived from one or other locality on the river banks, and is deposited further up or down the river, as the case may be, according to the tide. In this manner the mud is continually moving from one spot to another, though in the aggregate the result is very trifling, and no great amount reaches the Humber. The sediment that does reach these waters by the rivers can only be a very small proportion of that which exists in the estuary ; and as it apparently does not come down the rivers flowing into the Humber, it must come in from the sea.

Mr. H. M. Platnauer, of York, has made some observations which are very interesting in connection with this enquiry. His method of procedure was very simple : 102 different samples were taken from the River Ouse, and a litre of each evaporated to complete dryness. The residue thus obtained was carefully weighed. An estimate of the quantity of water passing a given point in a given time was then obtained, and a short calculation gave the amount of solid matter passing York in any given period. Mr. Platnauer estimated that, taking everything into consideration, about 300,000 tons of material are annually carried past the city of York, in the waters of the Ouse. Unfortunately, no attempt was made to differentiate between the amount of solid matter in solution, and that in suspension.

What at first seems remarkable, however, is that when the water is swollen with flood, and, consequently, turbid, there is *less solid matter per litre* after evaporation than when the water is clear, and flowing under ordinary conditions. This is no doubt rightly accounted for by Mr. Platnauer, who states that in the former instance the water is principally surface water, while in the latter it has passed through the soil and rock, and reached the Ouse by means of springs. It is consequently then charged with much mineral matter.

If, therefore, during the brief periods of " fresh," when the water is full of numerous fine particles in suspension, there is less solid matter per litre than when the stream is normal and clear, it is evident that only a very small proportion of the quantity annually swept past York is in *suspension*. Of course, it is not forgotten that when the river is in flood a far greater quantity of water, and consequently solid matter, passes York in a day than when the water is low. But the period of flood is not a long one, and, as already suggested, much of the matter in suspension may never reach the Humber.

In this connection I should like to quote a paragraph from a paper by Dr. H. F. Parsons.* Referring to

* The Alluvial Strata of the Lower Ouse Valley, " Proc. Yorks. Geol. Soc.", 1877, pp. 214-238.

the Humber warp, he writes :—" The appearance and
physical characters of the warp are very different from
those of the sediments deposited by the rivers in the
non-tidal portion of their course, that of the Ouse being a
coarse, brown, sandy loam ; of the Aire, a black, loose,
woolly-looking earth. The warp, too, is most abundant
at high-water at Spring tides, and in dry weather ; and
least so at low-water, during neap tides, and when much
fresh water is coming down the rivers. These facts
seem to show that it has a different origin from the ordinary detritus carried down by rivers. It appears to
reach its maximum in the lower reaches of the Ouse,
probably because the opposing tendencies of the tide
to wash it up, and of the river current to carry it down,
are there most counterbalanced. On the hypothesis
that the warp is brought up by the tide from below, the
fact that it occurs higher up the rivers than the salt water
reaches, may be explained by supposing it to be washed
up, bit by bit, at consecutive tides, the part deposited
at one tide being carried up a little higher and again
deposited at the next. The velocity of the flow of the
tide being much greater than that of the ebb, its carrying
power would be proportionately larger. On the other
hand, the sea-salt being in solution, would be completely
washed away by the water coming down from above."

Earlier than this, James Oldham in 1853 read a
paper to the British Association (which met in Hull in
that year), in which he said —" From the best observations I have been able to make, I find that the deposit

does not take place either at the flood or at the ebb tide, or yet at any time when the water is in motion ; but only at high water, when it is in a quiescent state, and the quantity left is just in proportion to the depth of the tide at the time. Now, if the deposit be brought down the river, the only quiescent state it could have when so brought down, would be at the turn of the tide at low water, and therefore no accumulation could take place such as we have been describing, at least from that direction ; for immediately the current begins to fall with the flood, the whole of the loose deposit is again set in motion.

Taking, therefore, Sunk Island as the point for consideration, at the time of high water of spring tides, where is it likely the mud could come from, which is found in suspension above the surface of the land ? We have seen that the ebb could not deposit it, because of the current and the lowness of the surface of the water. Then finding that the deposit does take place, and can only take place at high water, if it does not come from the sea, whence can it come ? " This point is rather an important one.

Let us look outside the estuary for a moment. The waters of the North Sea are continually washing particles of rock, sand and mud in a southerly direction, and slowly but surely the material on the Yorkshire coast is travelling southward. It never travels in a northerly direction. A good illustration of this can be found in the chalk boulders around Flamborough Headland. As is well-known, the beach around that promontory is strewn with masses of chalk of all sizes which have been dislodged from the cliffs. These can be seen in plenty in Bridlington Bay and further south, though naturally getting less plentiful as they get towards the Humber. Practically no chalk boulders are to be found north of the headland. This goes to prove conclusively, if proof be needed, that the beach material travels to the south ward.

The cliffs of the Holderness coast, made up entirely of soft glacial clays, vary in height from ten feet to fifty feet, and at Dimlington reach over one hundred feet. It has already been amply proved in these pages, both by antiquarian evidence and by direct measurements, that the whole of the forty miles of cliffs from Bridlington to Spurn are being eroded at an average rate of over seven feet per annum. A walk along the shore at any point, especially in the spring time, will convince the observer that this estimate must be fairly accurate, or at any rate not over estimated. As, therefore, the boulder-clay cliffs are being quickly washed away, the whole of the material must be gradually, or, as it is in some cases, quickly, converted into gravel, sand or mud, and carried southward.

It is generally admitted that a large quantity of the material is carried past the Humber mouth and is gradually silting up in the Wash and off the Lincolnshire coast, but at the same time a deal of the material must be brought into the Humber at each tide ; and when the winds are the strongest, and the rate of erosion is consequently the most severe, the inrush of water into the Humber is likewise the greatest. This water brings with it the cliffs in a modified form.

It would appear, therefore, that it is from the coast that much of the material in suspension in the Humber waters is derived, Of course, it does not follow that the mud now in the Humber is the result of one or two tides. The particles in the water may have been accumulating during several months, and undoubtedly pass and re-pass a particular point several times a week. Consequently, when the rivers flowing into the Humber are swollen with flood waters, and are swift, the muddiness observed near the entrances to the estuary is not necessarily due entirely to the additional material which they have brought down, but is more likely to be owing to the sediment in the Humber being stirred up.

A short time ago, during an exceptionally dry summer, the Humber water was, comparatively speaking, in a clear condition. This was not altogether due to the lack of material brought down by the rivers on account of their almost dry condition ; rather may it be attributable to the fact that the mud was not stirred up to any extent, but was brought in by the Humber, deposited at high water, and allowed to remain.

In the rivers Ouse and Hull, etc., as far as they are tidal, marine foraminifera and diatoms occur, and even marine fishes ; these have found their way up the rivers with the tides, and there is sufficient salt in the water for them to live. This being so, surely particles of mud can come into the Humber and up the rivers ?

Similarly, in various parts of the Humber, as at Brough, South Ferriby, and even as far as Goole, and up the Trent as far as Burton Stather, large pieces of the hard scum from the slag furnaces at Middlesbrough are regularly picked up, and put upon " rockeries." These are full of cavities containing gases, which enable the pieces to float. They also occur all along the south York-shire coast, and can be traced point by point, from say Bridlington to Hornsea, Withernsea, Spurn, Sunk Island, Paull, Hessle, and on to Goole. They have clearly been washed into the Humber, and how could they do this, and leave the mud behind ?

It will be remembered that a little while ago, a stranded whale, " serene and high," found its way into the estuary. Notwithstanding frequent attempts to allow it to drift out with the tide, it still came back. Finally a friendly ship towed it out to the North Sea, and cast it loose. And still it returned ; fortunately, on this occasion, being stranded near a fish manure works at Killingholme, where it was put to good use. If a dead whale could not be kept out of the estuary, how is it possible to make us believe (as some endeavour to) that no fine particles of mud can reach the Humber from outside ?

. . The work being done by geologists in connection with the deposits at the mouths of other rivers, suggests that a microscopic examination of the particles in the Humber water would probably enable it to be ascertained definitely whether the bulk of the material came from the cliffs of Holderness, or from the rivers Ouse, Trent and Hull. It would be interesting if such work were carried out, as was suggested by Mr. A. Harker in " The Naturalist " for February, 1912.

It must be borne in mind that the erosive action of the rivers is the greatest where the ground is the highest, and the gradient the steepest, viz., at their sources. It is here that the current is the swiftest, and consequently the rocks and soil on either side of and in the bed of the streams are dislodged and carried away. Very little erosion is going on near the mouths of the rivers. It naturally follows, therefore, that such particles as are brought down have come from their various tributaries, which stretch all through the north, south and west of Yorkshire, and, in the case of the Trent, much of the material is brought from the Midlands, and even from the far side of the Pennine Chain.

The rocks found in the boulder clays, though, of course, only a small proportion of the bulk of the cliffs of Holderness, are partly similar to those which the rivers at their sources pass through ; but the Holderness clays contain many minerals not in inland deposits, though it must not be forgotten that the rivers traverse boulder clay areas.

It would therefore seem that, although a fair amount of sediment may be carried into the estuary by means of the rivers, much of it is derived from the Holderness coast.

I do not go so far as to say, however, that anything substantial can be done to prevent the mud from silting up in the Humber. Groynes doubtless stay the effects of the sea in the vicinity of the places where they are erected. Away from the groynes, however, erosion seems to take place with renewed vigour. A suggestion has been made that without the erection of groynes the land at Spurn would be cut across, the Humber would change its channel ; the ports might be silted up, and their trade for ever be stopped! This alarming state of affairs (which, however, was not original, but had been set forth by James Oldham, so long ago as 1853), had little effect, especially as the engineering authorities responsible for Spurn Head were quite satisfied as to its security. The groynes have not been erected, the cliffs are still being eroded, and the Humber still flows seaward in its old, old way!

At the meeting of the British Association at Glasgow in 1901, Mr. W. H. Wheeler read some notes on the source of the warp in the Humber, printed copies of which were distributed at the time. Mr. Wheeler has evidently gone very carefully and thoroughly into the question, and, while there is much in the paper that I cannot but agree with, his conclusions are quite opposite to my own. In his opinion, none of the Humber detritus is, or could be, derived from the waste of the Holderness coast. He also makes the following somewhat remarkable statement :—

" *The physical features of the Humber in no way lend*

themselves to the statement that the water entering the Humber with the flood tide is charged with solid matter derived from the erosion of the Yorkshire cliffs, and that this solid matter is carried up the river to Hull and Goole. Apart from this, *it would be opposed to the laws of nature if this were the case."* If it is true that we *are* opposing the laws of nature then the question is a serious one for us!

Mr. Wheeler concludes that " It is evident that it is physically impossible for the mud in the Humber to be supplied from the waste of the Yorkshire coast." * This statement reminds one of the old story of the man who was unfortunate enough to find himself in the stocks. A legal friend, who happened to be passing, enquired the reason, and on being informed, emphatically declared that " they can't do it, sir ; you cannot be put in the stocks for that, sir. It's impossible, sir, it can't be done." " It's all right you preaching," was the reply, *" but I'm in ! "*

---

* These statements are repeated in Mr. Wheeler's " The Sea Coast," 1902. Evidently Mr. Wheeler himself, as time has gone on, has modified his views ; as in his evidence given before the Royal Commission on Coast Erosion on November 9th, 1906 (see " Report," Vol. I., pt. 2, 1907, p. 153), on being cross-questioned as to 969,000 cubic yards per annum of alluvial matter, washed from the Holderness coast, which " is apparently lost sight of," he states, " yes, *it is possible* a little of that goes into the Humber and has been deposited on Sunk island.' Thus what was " physically impossible " in 1902, became possible in 1906.

CHAPTER XXVI

NATURAL HISTORY OF THE RIDING—MAMMALS—BIRDS—
REPTILES AND AMPHIBIANS — FISHES — INSECTS —
MOLLUSCS—PLANTS.

THERE is ample geological evidence to prove that at
one time the whole of Great Britain was a part of the
Continent, the severance therefrom occurring at a comparatively recent date.   In this way can be explained the
remains of many strange animals found in the more recent
geological beds, just referred to, which have proved to be
particularly plentiful in East Yorkshire.

Through the unusually complete work of various
scientific societies in the riding, which have issued
a valuable number of monographs and other publications,
there is much more material available with regard to the
natural history of this district than is generally the case.
Owing to the exceptional geographical conditions also,
portions of East Yorkshire are of more than ordinary
natural history interest.

Of Mammals, few species now exist in East Yorkshire,
in common with the rest of the country.   Foxes are
abundant, and afford considerable sport.   The Wolds
are well-known for the large numbers of hares and
rabbits which they harbour ; though the last-named
are not nearly so common as they were when the great
warrens existed.   The badger is still in the district, and

the otter is not uncommon in the rivers and streams. Of the smaller species, the various kinds of vole, mouse, bat, stoat, etc., occur in numbers, which vary according to the suitability of the situation. Perhaps, generally speaking, we can say that East Yorkshire has a rather unusual number of Mammals within its area. With

A YOUNG WHALE AT KILNSEA.

regard to the larger forms, such as the roebuck, red deer, etc., these only now occur preserved in the parks. There was formerly a herd of wild white cattle at Burton Constable.

Having a sea-board, East Yorkshire can include amongst its fauna the whales, dolphins, seals, etc., various species of which are washed up from time to time. The

253

more important occurrences are the type specimen of Sibbald's Rorqual, washed up at Spurn in 1836, the skeleton of which is now in the Hull museum ; and the large sperm whale, washed ashore at Tunstall in 1825, which was the cause of an expensive lawsuit. In this the lord of the manor won his case, and the skeleton is now in the park at Burton Constable.

East Yorkshire has long been known as a paradise for the bird-lover. On Flamborough Headland, and par-

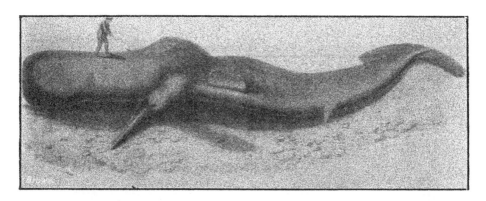

SPERM WHALE STRANDED AT TUNSTALL IN 1825.
(From an old print in the Museum, Hull).

ticularly on its northern half, where the cliffs rise in a perpendicular wall, innumerable sea-fowl regularly breed, and in the summer time the rocky ledges are occupied by myriads of guillemots, puffins, and razorbills. Other birds also live on the cliffs ; the peregrine falcon, which has recently taken up its quarters again after an absence of nearly 30 years, being a great attraction. An area of even still greater interest occurs on Spurn Point, one of the few localities in Great Britain where the lesser tern

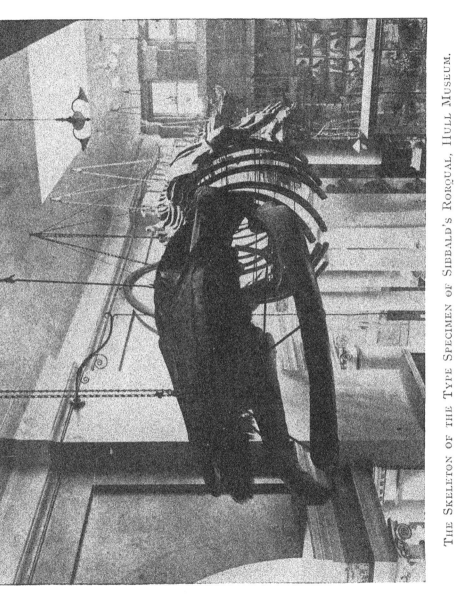

THE SKELETON OF THE TYPE SPECIMEN OF SIBBALD'S RORQUAL, HULL MUSEUM.

breeds in numbers. This species, together with the ringed plover (and recently the oyster catcher), lays its eggs on the sands, where they are well protected in the breeding season, by paid watchers. The Wold country, formerly the breeding place of the great bustard and dotterel, still harbours the stone curlew and other rarities. On the moors to the south-west of the riding the black-

[Photo by]   [Oxley Grabham.
OYSTER CATCHER ON NEST AT SPURN.

headed gull, curlew, etc., breed, and within the area is one of the most northerly localities for the nightingale.

At Hornsea Mere is a heronry, and on the islands and around the side of this lake the great-crested grebe, mallard, shoveller, pochard, water rail, and coot regularly nest. Recently the bearded tit has been introduced.

The moors and sandy wastes are inhabited by lizards, grass-snakes, and other harmless species. On some of the moors the viper still exists, but it is not common. In

Photo by]   NEST OF LITTLE TERN AT SPURN.   [*R. Fortune.*

Photo by]   NEST OF RINGED PLOVER AT SPURN.   [*R. Fortune.*

R

the immediate vicinity of Hull the grass-snake is particularly abundant, specimens over 3 feet in length being quite frequent. The marshy nature of certain parts of Holderness results in the presence of innumerable

"EGG-CLIMMER" AND HIS OUTFIT, BEMPTON.

streams and ponds, in which frogs, toads, and newts abound.

The streams issuing from the Wolds, particularly in the vicinity of Driffield, and many of the tributaries of the Derwent, are famous for their fish, and are largely

Heronry at Hornsea.

resorted to by anglers ; the Driffield streams being noted for their fine trout. The River Hull, in addition to its trout and pike, is inhabited by perch, miller's thumb, the three- and ten-spined stickleback, gudgeon, roach, chub, dace, bream, minnow, loach, grayling, burbot, broad and sharp-nosed eel, sea lamprey, lampern, and flounder. Some of these only occur in the tidal parts of the river.

Hornsea Mere is a favourite fishing ground, and perhaps is principally interesting from the extraordinary catches of pike, examples weighing from 20 to 24 lbs. each being not at all uncommon.

The sea on the east of Yorkshire abounds in marine fish, where, particularly in Bridlington Bay, various species are very plentiful, and afford sport to thousands of summer visitors, besides providing a livelihood to many fishermen throughout the year. Occasionally rarer forms are secured, such as the opah, the fishing frog, etc. The dog-fish and other smaller sharks also occur.

The common lands in the south and west of the riding, the high Wolds, the Holderness flats, and the sandy seaboard, are each inhabited by their characteristic species of insects, and many kinds occur which are scarce or absent in other parts of the country. The Spurn district particularly is the home of many interesting butterflies, moths, beetles, etc. In certain seasons the death's head hawk moth, the small elephant hawk moth, and other interesting forms are plentiful.

The quantity of timber imported at Hull from the

Photo by   [E. W. Wade.

YOUNG PEREGRINE AT BEMPTON.

Baltic has resulted in many continental insects being introduced. Perhaps the timberman (*Acanthocinus aedilis*), with its extraordinary antennæ, several times the length of its body, is the most remarkable.

As in the case of the insect fauna, East Yorkshire is a particularly favourable district for various species of mollusca, which in parts literally abound in thousands. The Spurn district especially swarms with innumerable

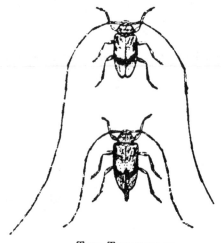

THE TIMBERMAN.
(*Acanthocinus aedilis*).

varieties of the land snail, *Helix nemoralis*, while on certain sections of the cliffs north and south of Bridlington the small-banded *Helix virgata* occurs in extraordinary abundance. The numerous canals, drains, becks, and ponds, also yield various species of fresh-water shells, and in Hornsea Mere and in the waters in some of the private parks, the swan mussel (*Anodonta cygnaea*) attains enormous proportions.

Photo by]   NEST OF OYSTER-CATCHER AT SPURN.   [*Oxley Grebham.*

GUILLEMOTS AT BEMPTON.

Though the coast-line of the East Riding is not more than about 40 miles in length, it provides exceptional facilities for the investigation of the marine life. The rocky ridge known as Filey Brig is the home of innumerable rare forms which can be easily collected at low spring-tides. The rocky pools in the vicinity of Flamborough also provide the home for many interesting shells, and in all directions the chalk scars are bored by the rock-boring species, *Pholas* and *Saxicava*. South of Bridlington large razor-shells, and the massive bivalve, *Cyprina islandica*, thrive in great numbers, while the mud-flats near Spurn are inhabited by the delicate shelled bivalve, *Scrobicularia piperata*. Within the estuary are innumerable varieties of *Tellina*, *Cardium*, *Mytilus*, etc. These occur in such enormous numbers that occasionally after a heavy tide they are banked up on the beach to a depth of as much as 2 feet.

The brackish waters on the shores of the estuary are occupied by various species of *Paludestrina*, one of which, *P. jenkinsii*, has extended its range in recent years to a very large extent. It has also apparently modified its habitat, and can now be found in almost pure fresh water.

Botanically the East Riding may be divided into three distinct areas, each with a characteristic flora. There is the Plain of Holderness with its old lake beds, in which are the sub-fossil remains of the arctic birch, scots fir, bird cherry, oak, hazel, iris, marsh pea, etc. The carr-lands and boggy grounds are the home of typical marsh-loving species, such as the bog bean, marsh pea, and grass of parnassus, while near the coast are the sea-side sedge, sea flote grass, marram grass, michaelmas daisy, sea lavender, wormwood and glasswort. The Spurn peninsula is the only locality in the riding, and one of the few localities on the east coast, where the beautiful sea-holly flourishes.

Here also may be seen the sea buckthorn, convolvulus, and other species, all of which help to bind the sand together. The pools on the Humber side of Spurn are inhabited by one of the very few species of marine flowering plants, namely *Zostera marina*.

The Wold area was formerly remarkable for its great growth of oak and other timber, and to-day parts of it are well-wooded with beech, oak, and ash.

The low-lying land to the west of the Wolds includes tracts of aboriginal common, where heather thrives, and where there are whole acres of cotton grass, rose bay willow-herb, marsh gentian, and similar species.

The estuary of the Humber also, with its large stretches of mud-flats, has a flora of its own ; the dominant plants being members of the Goosefoot family.

In the springtime the Holderness drains are resplendent with the marsh marigold ; later in the year with the yellow water-lily and the delicate " flowering rush " ; while in the autumn the banks are brilliant with masses of the purple loosestrife.

In the immediate neighbourhood of Hull the aspect of the flora has considerably changed, more particularly in recent years, by the introduction of a large number of foreign plants or " aliens." The seeds have been introduced in merchandise, and have been distributed throughout the district in a variety of ways.

At one time teazels were largely cultivated, and were sent into the West Riding in connection with the woollen industries. Mustard also was extensively grown in the vicinity of Sunk Island, but had to be discontinued on account of the ravages of the " mustard bug."

The Land of Green Ginger, a well-known business locality in the heart of the city of Hull, reminds us of the time when the district was occupied by gardens in which various pot-herbs flourished.

CHAPTER XXVII

EXTINCT ANIMALS: PRE-GLACIAL, GLACIAL AND POST-GLACIAL.

PARTLY as the result of various engineering undertakings, but principally from the inroads of the sea along the coast, the structure of East Yorkshire has been fairly well ascertained. Incidentally it has been possible to gather quite a large series of bones, teeth, etc., of animals which once existed in the district. These occur in various ways. Sometimes they can be found in old sand-dunes, where they have been left by the hyaenas or other carnivorous animals. At times they are found singly in the boulder-clay, where they have been left as " erratics " in the ice ; they may exist in a more or less water-worn condition in the gravel-mounds ; or the whole skeleton of an animal may be found, just as it had died, centuries ago, in a peat-bog.

With regard to age, the extinct animals of East Yorkshire may be placed under three heads : (1) Pre-Glacial, (2) Glacial, (3) Post-Glacial. As might be expected the earliest deposit contains a great proportion of remains of animals very different from those with which we are now familiar, while the most recent bed includes remains of animals approaching those living in the district to-day.

The Pre-Glacial remains occur on the old shore-line buried beneath the Glacial Beds. They have been found at Sewerby and Hessle. Many of them bore the marks of the teeth of hyaenas. The list includes the mammoth (*Elephas primigenius*), the straight-tusked elephant (*Elephas antiquus*), rhinoceros (*Rhinoceros leptorhinus*), hippopotamus (*Hippopotamus amphibius*), the horse, the Irish elk (*Cervus megaceros*), bison, and water vole. In addition

TOOTH OF *Rhinoceros tichorhinus* (actual size) FROM THE BEACH AT DIMLINGTON.

are remains of birds, a snake, codfish, and some shells From the old beach at Hessle have been obtained remains of the mammoth, rhinoceros, horse, red deer, and reindeer.

The Boulder Clay and the mounds of glacial gravel contain the remains which have been classified under the heading " Glacial." In both deposits the relics occur as

" erratics," but are more water-worn in the gravel beds. Among the species identified in the gravels, principally at Burstwick and Kelsey Hill, are the mammoth, straight-tusked elephant, Irish elk, reindeer, red deer, bison (*B. priscus*), Bos (*B. primigenius*), ox, horse, rhinoceros (*R. leptorhinus*) and walrus. The teeth-marks on some of these probably indicate the presence of hyaenæ. The Boulder Clay proper has yielded rhinoceros (*R. ticho-rhinus*), and mammoth.

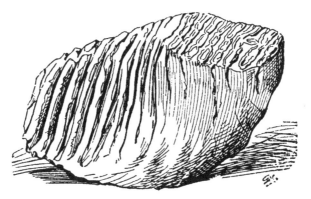

MAMMOTH TOOTH, WEIGHING 9½ LBS., FOUND AT WITHERNSEA.

Another bone-bearing bed, though of a different nature, and occupying the summit of a hill a hundred feet in height, occurs at Elloughton, near Brough. The remains found there include those of mammoth, straight-tusked elephant, red deer, bison, ox, and horse.

A further interesting deposit, evidently occupying the site of an old lake-bed, occurs at Bielsbeck, near Market Weighton, and is probably of Glacial age. It was first described at the beginning of last century, and more recently has been excavated by the aid of a grant from the British Association. It has yielded remains of mammoth, lion, rhinoceros, bison, *Bos primigenius*, red deer, Irish elk, horse, dog or wolf, and duck.

Of Post-Glacial age are the old lake-beds which occur in hollows on the Drift. These are best examined on the coast sections, though they are occasionally excavated by agricultural and other operations inland. The remains found therein include those of *Bos primigenius*, *Bos longifrons*, Irish elk, red deer, reindeer, horse, beaver, dog or wolf, birds (including duck), perch, and pike. In addition, there is a large list of fresh-water Mollusca, principally, however, representing those species still existing in the district.

CHAPTER XXVIII

EAST YORKSHIRE : ITS PEOPLE, RACE, DIALECTS, SETTLE-
MENTS, POPULATION—BRITONS—ROMANS—SAXONS
—DANES—HISTORY.

IT must not be thought that prior to the Roman
Invasion the occupants of these islands were of the semi-
savage type which some historians seem to infer. From
the writings of the Romans themselves, and also from
unwritten, but, nevertheless, unquestionable records,

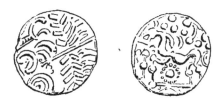

BRITISH GOLD STATER FOUND AT HORNSEA.

it is clear that before the Roman soldiers trod our soil,
the people in East Yorkshire were well versed in government and warfare, had both a gold and silver coinage,
and, judging from a recent discovery made on the Humber
shore, it is evident that they were able to equal the
more civilised people of the twentieth century in the
way in which they issued gilded base coin in place of
the gold staters which were then circulated.

East Yorkshire was once occupied by the Brigantes.

According to Ptolemy's Geography, which was prepared about the year 120 A.D., there was in his time a tribe known by the name of the Parisi, from which it would seem that they were a branch of the Parisii on the Seine, who have left their name in the present city of Paris. Petuaria, the town of the Parisi, has not yet been definitely located ; though probably it was at Brough on the Humber, which conforms with Ptolemy's measurements, and has yielded unmistakable evidence of both British and Roman occupation.

As to when the Parisii arrived in this country there is at present no evidence. It is not known whether they were here before or after the Brigantes, and, if the latter, whether they succeeded in claiming a part of the territory by main force. Neither is it known when they first came into East Yorkshire. It is known that they were independent of their powerful neighbours, and that they had a coinage of their own.

Of the Roman occupation we have many important evidences in the riding. It is not apparent, however, that these people have left any racial traces, though there are a few words still existing which have a Roman origin. The withdrawal of the Roman troops from these parts in the fourth century seems to have been effectively carried out. Immediately following this there is a distinct break in our history, and the next event of which we have any record was the Anglo-Saxon Invasion.

Respecting the Angles, Saxons, and Jutes, we have some very definite information of a reliable character.

The Angles came from Angleland, situated in a district now known as Schleswig, in the centre of the peninsula between the Baltic and the northern seas. These are practically the people from which the population of East Yorkshire has descended. North of Angleland lived a kindred tribe, the Jutes ; and, to the south, were the Saxons. During the fifth century Angle, Saxon, and Jute, were, according to Green, " being drawn together by the ties of a common blood, common speech, common

ANGLO-SAXON CINERARY URNS FROM SANCTON.

social and political institutions. Each of them was destined to share in the conquest of the land in which we live ; and it is from the union of all of them when its conquest was complete that the English people has sprung."

These are the people who for so long have been known by the not quite correct name of Anglo-Saxons. In East Yorkshire there are many traces of the Anglian invasion and the subsequent colonisation. Several of the existing villages occupy precisely the sites of the first clearings made by the Anglo-Saxons. The names of the villages, in some instances slightly modified, are those which were given by these people ; our local dialect contains numerous proofs of Anglo-Saxon origin or influence. In the churches, and in the numerous relics dug up from time to time, there is unimpeachable testimony of the arts and religion of these our forefathers.

In various burial places in the riding innumerable weapons, implements, utensils, and ornaments of unquestionable Saxon origin have been unearthed ; and in many places within the district entire Anglian cemeteries have been discovered. Of this period also there is an interesting reference to Goodmanham in the Venerable Bede's " Ecclesiastical History," a work prepared early in the eighth century. In this he clearly shows that a heathen temple once existed at Goodmanham, and on its destruction a Christian temple was set up in its place. Numerous words of Danish and Anglian origin are likewise still in use.

The Danish villages, which are so frequent in the East Riding of Yorkshire, as well as in North Lincolnshire, can be at once identified by the suffix " by," " thorpe," or " thwaite." The site of a Danish village was usually chosen midway between two of the older Anglian settlements : it was placed upon the side of an existing road, and it was sometimes carved from the Anglian lands on either side. This is strikingly shown when we come to examine the positions of the Danish villages in the district. Such wayside settlements are Carnaby and Bessingby, on the road from Bridlington to Driffield ; Lund and Tibthorpe, on the road from Wetwang to Beverley ; and Willerby on the road from Hessle to Beverley. When, as was sometimes the case, the new settlement was planned at a little distance from the existing road, a new road, running at right-angles from the old one, and leading directly to the settlement, was formed. Skidby, Towthorpe, Kirby Grindalythe, and many other villages are instances in point.

By taking the areas of Danish and Anglian settlements respectively, it is clear that the greater part of the country was already taken up by the English, and that consequently only small areas were available for the newer Danish settlers.

In parts of East Yorkshire, particularly at Flamborough and some of the more inaccessible areas of the Wolds, the influence of the Norse is particularly marked in the dialect, many of the words still used being almost pure Scandinavian.

The first historical reference to this district, appears to be a brief notice of the Parisi and of Petuaria in Ptolemy's Geography, which has already been referred to. Of the district in Roman times we have many evidences in the form of relics ; but it cannot with any certainty be stated that a single Roman station has been definitely located within the area.

In the early days of Christianity, Nunburnholme, Watton, Beverley, Bishop Burton, and Weaverthorpe played their parts. There was a Saxon nunnery at Watton, and a Saxon monastery at Beverley. St. John of Beverley, who was born at Harpham, had a great reputation, and the magnificent minster was built on the site of his shrine. At the battle of Stamford Bridge on the borders of the riding, Harold overthrew his brother Tostig. After the Norman Conquest, Drogo de Brevere received a portion of Holderness, and erected his castle on the mound at Skipsea, which was the first stronghold of the seignory of Holderness. Later the Earls of Albemarle moved the headquarters to Burstwick, and finally to Burton Constable. Generally speaking, however, the East Riding was apparently not so well fortified in Norman times as were the North and West.

In 1399, Henry Bolingbroke, and later, in 1471, Edward IV. landed at Ravenspurn.

In the Pilgrimage of Grace, East Yorkshire was to the fore, the first assembly of the rebels being at Weighton, under Robert Aske.

During the Civil War in 1642 the East Riding played a prominent part, inasmuch as Hull, which was then the stronghold of the north, was coveted by both Royalists and Parliamentarians. The people of Hull sided with the Parliament, and were able to withstand the brief siege which followed. Their action had much to do with the success of the Cromwellian party.

CHAPTER XXIX

ANTIQUITIES: PRE-HISTORIC—EARTHWORKS AND CULTI-
VATION TERRACES—BARROWS—RUDSTON MONOLITH
—ROMAN REMAINS—SAXON ANTIQUITIES.

FEW districts can compare with the East Riding of
Yorkshire with regard to its wealth of relics relating to
the people who occupied it in the past. The high
Wolds are not only dotted over by hundreds of barrows
or tumuli containing the remains of departed British
chiefs, but the whole area is ramified in all directions
by earthworks and fortifications, some being of gigantic
proportions. Even in the low ground of Holderness,
which in Pre-historic times would be largely an impassable
swamp, there are important pre-historic fortifications,
sometimes taking the form of large earthworks, at others
pile structures which were built within the margins of the
meres. The earthworks on the Wolds extend in more
or less straight lines, and usually consist of a large embankment adjoined by a deep ditch; though in some cases
there are two, three, four, five, or more earthworks and
ditches running parallel with each other. Perhaps the
most remarkable of these pre-historic structures is the
mis-named "Danes' Dyke," an enormous embankment
stretching from Sewerby near Bridlington, right across
Flamborough Headland to Cat Nab. Some miles to
the west of these are Argam Dykes, and further to the

west still, are other series, many of which, unfortunately,

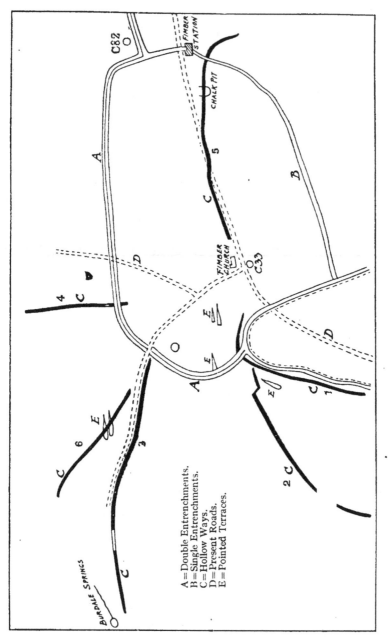

PLAN OF PREHISTORIC EARTHWORKS AT FIMBER.

A = Double Entrenchments.
B = Single Entrenchments.
C = Hollow Ways.
D = Present Roads.
E = Pointed Terraces.

are being gradually obliterated by agricultural operations,

though, happily, this cannot occur with the larger examples. Excavations have shown that these structures were erected during the Neolithic or New Stone Age.

Perhaps the most formidable earthwork is that of Skipsea Brough, near Hornsea, where there is a large central mound 70 feet high, and covering five acres of ground. At a distance of 220 yards from the central mound is a series of mounds and ditches extending for

NEOLITHIC HAMMERS FROM EAST YORKSHIRE.

half-a-mile in length, which occupy a somewhat semi-circular position, and in places are as high as the central mound. These elaborate earthworks protect the high ground only; the low land in British times being an impassable swamp. Of smaller size, but none the less interesting, are the " Castles " near Swine in Holderness, where there is a large oval central mound protected by a large encircling embankment and ditch.

279

LOST TOWNS OF THE YORKSHIRE COAST

At two or three places on the Wolds are the curious features described by the late J. R. Mortimer, as " habitation terraces." These usually consist of a series of enormous step-like terraces on the sides of the dry chalk valleys, a particularly fine example occurring at Croombe near Sledmere.

The barrows, of which nearly 400 have been excavated in the East Riding, vary in size ; some being of such enormous dimensions that they have been mapped as natural hills. These contain the remains of British

FLINT DAGGER FROM COTTINGHAM.

chiefs, which were either buried in stone cists, hollowed tree trunks, or merely deposited in holes cut into the solid chalk. Sometimes the remains were cremated and placed in large cinerary urns.

In both cases various relics were buried with the dead, and throw considerable light on the mode of living of these early occupants of the district. Among the objects are drinking cups and food vases of earthenware, and the curious smaller vessels, known as " incense cups " ; and weapons, implements, etc., such as daggers,

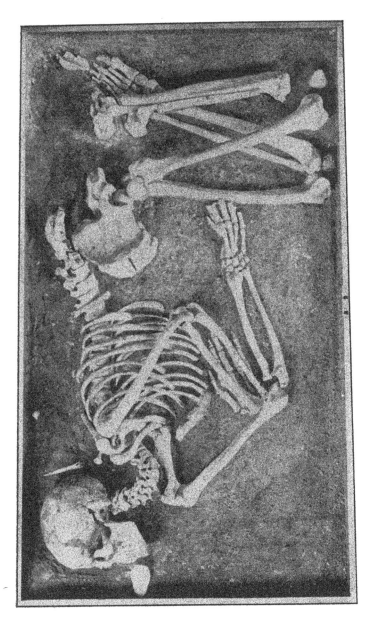

Skeleton of Ancient Briton from East Yorkshire, showing the characteristic crouched position.

(Note the hair-pin behind the head).

knives, beads, and pins. Most of these barrows belong either to the Neolithic or to the Bronze Age, and are circular in form. They usually occur singly, in prominent

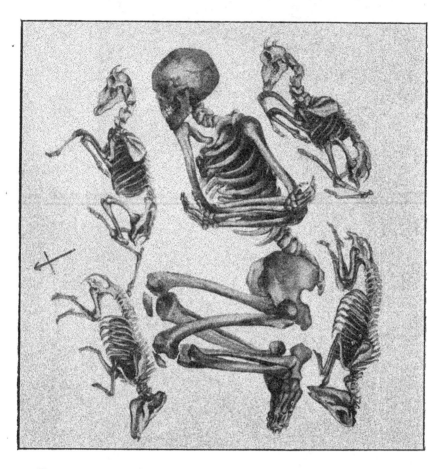

BURIAL OF THE EARLY IRON AGE IN ONE OF THE SO-CALLED "DANES' GRAVES" NEAR DRIFFIELD.

Accompanying the human skeleton are the remains of two pigs and two goats.

positions on the top of the Wolds. Of a later period, namely the early Iron Age, are other sets of burial mounds. These are usually much smaller in size, sometimes only

PORTION OF PAVEMENT FROM ROMAN VILLA AT HARPHAM, NOW IN THE HULL MUSEUM.

8 or 10 feet in diameter, and are generally crowded together. Such an assemblage occurs in the so-called "Danes' Graves," near Driffield, where some scores of small barrows are grouped closely together. These have yielded weapons and utensils of quite a different type from those already described, and among the relics found are two or three examples of chariots.

Near the church at Rudston is the one pre-historic monolith that exists in the riding, and it consists of coarse "moor grit," such as occurs on the moors above Robin Hood's Bay, some miles to the north.

Of Roman remains there are a great number in the riding ; though there are no large and important stations other than that of York (Eboracum), which is on the western border. Nevertheless many interesting evidences of Roman occupation have been revealed from time to time. At various places on the Wolds, remains of Roman villas, pavements, hypocausts, and other traces of occupation have been discovered ; while pottery, coins, etc., occur in great profusion in most parts of the riding. At Filey, at the top of Carr Naze, the remains of a Roman look-out post or lighthouse have been unearthed ; at Harpham near Driffield, the site of a villa was recently exposed ; at Brough on the Humber, which was the northern ferry on the road from Yorkshire to Lincolnshire, innumerable Roman relics have been found, and along the coastline between Bridlington and Spurn, Roman coins, pottery, etc., are constantly being washed out by the sea, and are some indication that a Roman road once existed along this coast-line.

Of Anglian or Anglo-Saxon relics the district has yielded a fair number. It is rather a remarkable fact that in several places large cemeteries have been discovered during agricultural operations, which occur on sites not occupied by villages to-day; and from the number of remains found it is clear that in Saxon times there were important communities in places now occupied by fields. Such cemeteries have been discovered near Market Weighton, Londesborough, Sancton, Newbald, Sledmere, Acklam, Driffield, Garton Slack, and Kilham. At other localities burials in smaller numbers have been recorded. There are two kinds of Saxon cemeteries, namely, those containing the bodies buried in the

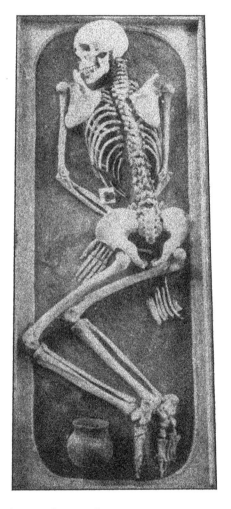

ANGLO-SAXON INTERMENT, SHOWING IRON BUCKLE, RIBS OF OX, ETC.

usual way, generally in rows, and accompanied by swords,

spears, shields, knives, daggers, jewellery, etc., and those which consist of cremated interments only. In the latter case the remains are placed in elaborately ornamented cinerary urns, which, as at Sancton, have been found in great numbers.

CHAPTER XXX

ARCHITECTURE : ECCLESIASTICAL, DOMESTIC — WOLD
CHURCHES—HOLDERNESS CHURCHES : HEDON, PAT-
RINGTON, BEVERLEY, HULL, ETC.—WRESSLE CASTLE
—BURTON AGNES AND BURTON CONSTABLE HALLS—
WILBERFORCE HOUSE—FARMHOUSES.

A PRELIMINARY word on the various styles of English
architecture is necessary before we consider the churches
and other important buildings of our district.

Pre-Norman or, as it is usually, though with no great
certainty termed " Saxon " building in England, was the
work of early craftsmen with an imperfect knowledge
of stone construction, who commonly used rough rubble
walls, no buttresses, small semicircular or triangular
arches, and square towers with what is called " long-and-
short work " at the quoins or corners. It survives almost
solely in portions of small churches.

The Norman Conquest started a widespread building
of massive churches and castles in the continental style,
called Romanesque, which in England has the name of
" Norman." They had walls of great thickness, semi-
circular vaults, round-headed doors and windows, and
lofty square towers.

From 1150 to 1200 the buildings became lighter, the
arches pointed, and there was perfected the science of

vaulting, by which the weight is brought upon piers and buttresses. This "Gothic" method of building originated from the endeavour to cover the widest and loftiest areas with the greatest economy of stone. The first English Gothic, called "Early English," from about 1180 to 1250, is characterised by slender piers (commonly of marble), lofty pointed vaults, and long, narrow, lancet-headed windows. After 1250 the windows became larger, divided up, and ornamented by patterns of tracery, while in the vaults the ribs were multiplied. The greatest elegance of English Gothic was reached from 1260 to 1290, at which date English sculpture was at its highest, and art in painting, coloured glass making, and general craftmanship at its zenith.

After 1300 the structure of stone buildings began to be overlaid with decoration, the window tracery and vault ribs were of intricate patterns, the pinnacles and spires loaded with ornament. This latter style is known as "Decorated," and came to an end with the Black Death, which stopped all building for a time.

With the changed conditions of life the type of building changed. With curious uniformity and rapidity the style called "Perpendicular," which is unknown abroad, developed after 1360 in all parts of England, and lasted with scarcely any change up to 1520. As its name implies, it is characterised by the perpendicular arrangement of the tracery and panels on walls and in windows ; and it is also distinguished by the flattened arches and square arrangement of the mouldings over them, by the elaborate vault traceries (especially fan vaulting), and by the use of flat roofs, and towers without spires.

The mediæval styles in England ended with the dissolution of the monasteries (1530-1540), for the Reformation checked the building of churches. There succeeded the building of manor-houses, in which the style called " Tudor " arose, distinguished by flat-headed windows, level ceilings, and panelled rooms. The ornaments of classic style were introduced under the influences of Renaissance sculpture and distinguish the " Jacobean " style. About this time the professional architect arose ; hitherto, building had been entirely in the hands of the builder and craftsman.

When the early builders wanted a larger place of worship they did not pull down all the work of their forefathers, but added to or altered it. Consequently many of their buildings exhibit the workmanship of several periods.

While there have been important abbeys and monasteries at Meaux, Watton, Hull, Haltemprice, Nunkeeling, Nunburnholme, etc., very few traces now remain ; the principal ones being the remains of the Augustinian Priory at Bridlington, the nave of which is in use to-day as a church ; the small Cistercian nunnery at Swine, the choir of which is still used as a parish church ; and Kirkham Abbey. The ruins at Kirkham represent all that is left of an important Augustinian Priory. There are also smaller remains at other places.

When, however, we come to examine the churches in

the riding, we find the district remarkably rich. At Holy Trinity, Hull; at Hedon, Patrington, St. Mary's (Beverley), and Cottingham, are parish churches of considerable importance. In addition are the magnificent collegiate structures of Howden and Hemingborough, and above all Beverley Minster. Saxon work is exhibited at Skipworth, Wharram-le-Street, Nunburnholme, Weaverthorpe, Patrington, and Aldborough. At the last three places the Saxon work is represented by cross-shafts and sundials respectively. The Wold churches are particularly rich in Norman work; and chancel arches, tympana and fonts, showing various phases of Norman workmanship, occur in quite a large number of them.

Perhaps the most beautiful example of ecclesiastical architecture in Yorkshire is Beverley Minster. In the beauty of its proportions and in the grace of its towers and flying buttresses it certainly exceeds any other building in the riding. It exhibits admirable examples of the Early English and Perpendicular styles. The Minster also possesses the relics of St. John of Beverley, and the well-known Percy Tomb.

Holy Trinity Church, Hull, the fifth largest parish church in England, is remarkable for the fourteenth century brickwork in the chancel and transepts. It bears many evidences of rough treatment at various periods, and has many features about it which are not favoured by enthusiastic students of architecture. The transepts and central tower are of the Early Decorated Period; the choir, Later Decorated; and the nave, Perpendicular. The building was restored by Sir Gilbert Scott about half-a-century ago.

Patrington Church, known as the " Queen of Holder ness," is one of exceptional beauty and grace, and is said to be one of the finest examples of fourteenth century architecture in England. It consists of a nave, with aisles, chancel, transept with aisles, and central tower and spire; the latter, with its flying buttresses and coronets, being exceptionally beautiful. In this church is an Easter sepulchre, one of the most perfect in the country.

The church at Hedon has a square tower which forms a prominent landmark in the district, and has not inaptly been termed the " King of Holderness.'" It consists of a nave and aisle, chancel transepts and tower. The choir and transepts are Early English, the nave Decorated, the central tower Perpendicular in style.

St. Mary's Church, Beverley—this beautiful church is a mixture of the Perpendicular and Decorated styles of architecture, and contains many interesting features, even rivalling its neighbour, the Minster, in many respects. It is cruciform in shape, with a massive central tower which half-a-century ago was thoroughly restored by Sir Gilbert Scott. It contains many remarkable examples of fine sculpturings.

At Wressle are the remains of the oldest example of domestic architecture in the riding; Wressle Castle being formerly one of the palatial residences of the Percys, Earls of Northumberland. The portion remaining consists of the south side of a quadrangle. The structure is built largely of stone, whereas most of the old examples of domestic architecture are principally of brick. An exceptionally fine instance is the Hall at Burton Agnes, the home of the Boynton family. This has been described as one of the finest Elizabethan houses in the north of England. The Hall is approached through a picturesque gatehouse with octagonal turrets, and bearing the arms of James I. Adjoining the Hall are the remains of a still earlier building, the basement of which is vaulted, and is evidently of twelfth century work.

At Burton Constable, the ancient seat of the Constable family, is a fine mansion, one of the largest houses of its kind in the county. It contains workmanship of various dates, but the two " fronts " are of the time of James or Charles I. The house contains some fine pictures and rare furniture.

Also of Elizabethan date is Wilberforce House in High Street, Hull, a typical example of a wealthy merchant's residence, and originally the home of the Lister family. Eventually it came into the hands of the Wilberforces, who about 150 years ago made extensive alterations and additions. In addition to the early panelled rooms, there are many fine examples of Georgian decoration. To-day the building is used as a public museum

and memorial to William Wilberforce, the slave emanci-
pator, who was born there in 1759.

Formerly the typical East Riding farmhouse consisted
of a long building with thick low walls and a thatched roof.
It was divided in the centre by a passage, on one side of
which lived the farmer and his family, while on the other
the horses and cattle were kept. Very few, however,
of these now remain. Other interesting buildings are
the tithe barns, formerly erected in the vicinity of the
churches, one of the best remaining examples being at
Easington, near Spurn, which is largely in its origina
state, with thatched roof and strong supporting timbers
(see page 115).

Along the coast, the churches and older houses are
largely built of the rounded boulders gathered from the
beach. Further inland, brick or stone was used ; perhaps
the most notable example being Beverley Minster, the
fine soft grey colour of which is due to the fact that it is
principally built from the local Oolitic limestone from
Newbald.

CHAPTER XXXI

COMMUNICATIONS—ANCIENT ROADS—RAILWAYS— CANALS.

IN East Yorkshire the system of roads is a very old one, and due to a variety of causes. In British times, when the Wolds were thickly populated, trackways extended along the brows of the hills, and on the sides of the valleys, connecting one village community with another. In many cases these roads or trackways were protected by an earthwork, sometimes of large dimensions, which enabled the inhabitants to travel from one point to another under cover. In some instances British roadways were so well made and connected such convenient points, that they were afterwards utilised by the Romans. Such a road exists between York and Brough on the Humber, from which point there was a ferry to Winteringham, where the road is continued in almost a straight line to Lincoln. The Lincolnshire portion of this road from Lindum (Lincoln) to Eboracum (York) is of Roman construction, whereas the Yorkshire section which, as pointed out, had previously connected two British settlements, is not as straight nor as well made as the typical Roman roads.

Principally for military purposes the Romans had roadways in different parts of the riding, and though the sites of some of these can still be traced, and in other places are indicated by such names as " The Street," they do not appear to have been of the substantial nature that they were in other parts of the country. York was the centre from which the roads then radiated; and perhaps the best example we have in the riding is that extending from York through Stamford Bridge and Fimber towards Filey, though its precise position at the eastern end is not quite clear. At Stamford Bridge a branch of this road extended to Malton, which was a Roman station. The road to Brough, already referred to, passed from York through Barmby Moor and Market Weighton. In Holderness the nature of the country in Roman times would be such that roads of the ordinary form would be well nigh impossible; though, as there was a Roman station at Kilnsea, and Roman remains have been found at Swine and Bishop Burton, it is possible that means of communication extended from Market Weighton to the coast on the east. Holderness itself has yielded exceedingly few Roman remains, though a number have been found on the coast, where the land is highest; and it is fair to assume that the Romans would have a roadway along the east coast, the site of which is now some distance out to sea. In mediæval times a number of new roads came into existence largely as a result of the pilgrimages to the shrine of St. John of Beverley. A glance at a map will show that roads reached Beverley from all parts of the compass, like the spokes of a wheel; whereas none go straight through it. In more recent times there were a few coach-roads in the riding ; though on account of the district being somewhat isolated and away from the manufacturing centres, these were not numerous. That from Hull to York was perhaps the most important.

The railway system of the East Riding has its centre in Hull, from which city branches extend to all parts of the county. The busiest line is the North Eastern Railway branch to Selby, for along this is transported most of the vast amount of merchandise to and from the great Yorkshire manufacturing towns, Leeds, Bradford, Halifax, and Huddersfield. The traffic on this line is an index of the great import and export trade carried on by these towns with all parts of the world. To cope with it the North Eastern Railway has no fewer than eight docks at Hull, with a total water area of 140 acres, and quay spaces of 270 acres. Another important branch line of this company goes to York and another to Malton ; while the popular holiday resorts of the Yorkshire coast are connected with Hull by a line which passes through Bridlington and Filey. Two shorter branch lines run from Hull to Withernsea and Hornsea respectively, two places much frequented by Hull holiday seekers.

The Hull and Barnsley Railway line runs for a great part of its length in the riding, on its way to the South Yorkshire coalfield. By this route enormous quantities of coal are brought to the Hull docks for the use of shipping, and for exportation to foreign countries.

The Great Central is the third of the railway companies connected with Hull. The actual railway, however, terminates at New Holland on the Lincolnshire side of the Humber, and communication is maintained with Hull by the means of a steam ferry-boat service.

The canals of the East Riding are small in number, and unimportant; this means of transit having been almost entirely superseded by the railways. The longest

Photo by] THE MARKET WEIGHTON CANAL. [*S. H. Smith.*

is that connecting Market Weighton with the Humber, a distance of 10 miles. The Market Weighton end of this is quite disused, the decaying lock gates appearing picturesque in a mantle of weeds, while the canal itself is choked with numerous water-plants. Small canals connect Beverley, Driffield, and Leven with the River

Hull, which river is put to much the same use as a canal. In the West Riding, however, the canal system is of great importance, and large quantities of merchandise find their way from the port of Hull to the great manufacturing centres by means of canal boats which are towed down the Humber to Goole.

England's canals were made in the eighteenth century, before the advent of the railway, and they would still afford a cheap means of transit, if both national and local public bodies realised the importance of their development.

Sometimes the East Riding canals, as in the case of that at Market Weighton, served the double purpose of transit and drainage of the surrounding fenny country.

CHAPTER XXXII

ADMINISTRATION — PARISHES — RURAL DISTRICTS —
URBAN DISTRICTS — BOROUGHS — ECCLESIASTICAL
DIVISIONS—POPULATION.

To trace in all its details the system of local and central
government of the East Riding of Yorkshire would be to
trace the history of our constitution from the time of the
Saxon invasions. The administration of a district is
always closely connected with the divisions of the land,
and so we find it here. The unit of government in early
Saxon times was the township, each with an independent
local court, and each sending its representatives to a
central court. Groups of townships formed wapentakes
or hundreds, two terms almost synonymous in their use.
The wapentake (Anglo-Saxon, waepen-getæc, weapon-
taking) occurs only in the Anglian districts ; Yorkshire,
Lincolnshire, Nottinghamshire, Derbyshire, Rutland, and
Leicestershire. North of this district the shires are
divided into wards, and to the south, into hundreds.
Yorkshire and Lincolnshire were divided into trithings or
ridings, and strangely enough in Domesday Book, we
find that while the other two ridings are divided into
wapentakes, the East Riding is divided into hundreds.
The wapentake in its system of administration corresponds precisely with the hundred, a fact demonstrated
by the ease with which the terms could be interchanged.

The hundreds of the East Riding were eighteen in number, and their names, North, South, and Middle Holderness; Driffield, Warter, Pocklington, etc. At the time of Kirkby's Inquest in 1284, however, it was di-divided into its present six wapentakes, Buckrose, Dickering, Harthill, Holderness, Howdenshire, and Ouse and Derwent. This division is now of little importance, except from an antiquarian point of view.

We may obtain some idea of the administration of the riding in early English times by considering the customs of our Teutonic forefathers as given by the Roman historian Tacitus in his "Germania." According to him the tribes were sub-divided into *pagi*, which seem to correspond with the English *hundreds*. The pagus, which some say was originally composed of a hundred families, supplied a hundred warriors to the army and a hundred assessors at the judicial court of the chief.

The hundred was further sub-divided into vici, or townships. The township thus formed the basis of the general organisation; the hundred, a collection of townships; the shire a collection of hundreds.

Townships might be free, or established upon the lands of lords, but in both cases the chief officer of the township was the reeve, elected on the one hand by the freeholders, and on the other probably by the lord. Townships constructed on the latter principle formed manors, in which the duties and privileges belonging to the freeholders in a free township were taken by the proprietor of the soil.

Each free township had a court, the dependent township having a counterpart in a manor court. Higher than this was the hundred court, over which presided the hundredmen, and on which each township of the hundred was represented by the Reeve and four other members. Above all was the shire court or Gemot. In most cases the shires seem to be coincident with the old sub-kingdoms, and over them was appointed a royal officer, as shire-reeve or sheriff, not appointed in those days by the people, but by the king just as to-day, whose judicial and fiscal functions he carried out. Such was the local government in early English times, a system which can still be traced in our modern institutions.

Central government was carried on through a national assembly or Gemot of the nation. This was called the Witan or wise men, or the Witana-Gemot, or assembly of the wise men. Every freeman had the right to attend this assembly, but naturally attendance was difficult to the great majority of them, and thus the power and functions of this court gradually lapsed into the hands of the king and the chief lords.

Although the adminstrative powers of the Witan were similar to our modern Houses of Parliament, the composition of the two cannot be said to be alike, for in early English times the principle of representation was not understood.

In theory, the power of the witan was the power of a free nation ; the modern assembly of the Houses of Parliament is so in fact. However, during the centuries, the one has evolved from the other ; but to trace the course of events leading up to this result would fill many volumes.

Each of the Yorkshire ridings has the status of a county. The central government is vested in a county council, and the chief officers are the Lord Lieutenant and the High Sheriff; the one may be said to represent the king, and the other the county.

The East Riding returns six members to Parliament, the three territorial divisions, Holderness, Buckrose, and Howdenshire, and the three divisions of the borough of Kingston-upon-Hull each electing one member.

The administration of the Poor Law is vested in ten Poor Law Unions, each comprising a large number of parishes.

Since the act of 1894 the East Riding is divided for local administrative purposes into Civil Parishes, Rural Districts, Urban Districts, Municipal Boroughs, and County Boroughs. The smaller parishes have a parish council, if the population is over 300, and a parish meeting if under. More populous districts have their Urban District Councils, and less populous districts, Rural District Councils. The district councils and parish councils represent the old town moots, the members being elected by the people to manage local affairs.

Some of the larger towns of East Yorkshire, however, have a different form of government. Beverley, Hedon. and Bridlington are Municipal Boroughs, that is, each is governed by a Corporation or Town Council, consisting of Mayor, Aldermen, and Councillors. Kingston-upon-

Hull is a County Borough, having the powers of a county council, being ruled by a Mayor and Corporation, and having magistrates and a police force distinct from that of the county.

In most parts of England, just as parishes were co-extensive with townships, so in olden times bishoprics were in a great degree co-extensive with the shires or ancient kingdoms. In Yorkshire, however, this is seldom the case, for a number of townships often form a single ecclesiastical parish. Seventy-eight of the 177 ecclesiastical parishes into which the riding is divided are of this composite character, although their boundaries are not always co-terminous with those of the groups of townships. The East Riding is a portion of the diocese of York, and is governed ecclesiastically by an archdeacon. The district is divided into twelve rural deaneries, Beverley, Bridlington, Buckrose, Harthill, Hedon, Hornsea, Howden Hull, Pocklington, Settrington, Weighton, Escrick. There are two suffragan bishops, one at Hull, and one at Beverley.

In 1901, the population of the riding was 384,997, an increase of 43,451 on the previous census of 1891. This is a greater increase than we find during the same period in the North Riding ; but this is largely due to the growth of Hull, though Bridlington shows an appreciable increase. If these two exceptions are made, the remainder of the district shows a slight decrease in the population.

The area, being largely agricultural, the population per square mile is only 323.5, whereas the average population per square mile in England and Wales is 558. Taking away the population of Hull it leaves an average of 89 per square mile, for the rest of the riding ; and as Driffield, Bridlington, and other small towns are included, it naturally means that there are great areas which are very scantily populated.

CHAPTER XXXIII

AGRICULTURE— AREA — CROPS — CATTLE — INDUS-
TRIES — MINERALS — FISHERIES — SHIPPING —
HULL IMPORTS AND EXPORTS—CLIMATE AND RAIN-
FALL.

THE area of the East Riding is 751,890 acres, of which
no fewer than 674,782 are under crops and grass ; the
permanent grass land occupying 224,000 acres. Of the
arable land 243,000 acres are under corn, principally
wheat, barley, and oats. Of the remaining 208,000 acres,
84,000 are under turnips, swedes, and mangolds, and
90,000 are under clover and grasses under rotation.
Smaller quantities of potatoes, rape, vetches, etc., are
grown, and there are small areas covered by fruit trees,
but these are of little importance. The figures of course
vary each year in consequence of the rotation of crops,
but they are drawn from the most recent statistics
available, and may be looked upon as fairly representative.

Flax was once much grown in different parts of the
riding ; and at Winestead and other places the receiving
buildings still exist, though it is a long time since the
beautiful flower was seen in the district.

Formerly caraway was grown on the Wolds ; but there
appears to have been no crop during the last quarter of a
century, though odd plants of it still appear in unexpected
quarters.

With regard to stock, sheep farming is extensively carried on, there being 500,000 sheep in the last returns. To a smaller extent horses, cattle, and pigs are reared, the respective figures being 40,000, 96,000, and 55,000 ; though with regard to the last-named there seems to be a falling off in the numbers recently, due possibly to the restrictions which have been enforced. Amongst the cattle are the " short-horned " variety for which Holderness has some reputation.

In view of the size of the East Riding, the industries and manufactures are comparatively small, being almost entirely centred in and around Hull. At some of the market towns, milling, farm implement making, and the manufacture of artificial manures are carried on ; and at Selby, Hessle, and Beverley a number of small ships are built. In recent years important flour mills, soap works, and gunpowder factories have sprung up at Selby.

In the ordinary sense of the word there are no mines in the East Riding, and minerals, as such, are not worked. This arises from the fact that geologically speaking the district is largely made up of the newer formations. At Hessle, and at various other parts of the Wolds, the chalk is quarried, and is used for ballast, for the manufacture of whiting and cement, and to an annually decreasing extent, for road repairs. In the north-west portion of the riding, the Oolitic limestone is burnt for what is known in the trade as " Lias lime." At Brough on the Humber, Burstwick and Kelsey Hills in Holderness, and to a lesser extent at other places, gravel and sand are excavated, being largely used in connection with dock works and for railway ballast.

On the old Humber silts along the Humber shore, and in the valley of the Hull, is admirable clay for brick-making, and that industry is carried on to a small extent at different points. More inland the boulder clay is occasionally used for brick and tile making, but on account of the impurities the results are not quite so satisfactory.

At Filey, Flamborough, Bridlington, Hornsea, and Withernsea, fishing is carried on to some extent. The fishing is generally from peculiar craft known as " cobles " and is mainly accomplished by means of lines, though drift nets for herrings and mackerel are employed, and ' pots ' for crabs and lobsters. At Filey a fair quantity of cod, ling, haddock, sole, plaice, and herring is obtained during the year ; though, as at the other places mentioned, the summer visitors occupy much of the fishermen's time. South of Filey Brig, salmon are secured by means of nets, during the summer time, this sport being keenly participated in by the visitors.

The value of the fish landed at Filey is nearly £4000 per annum. Flamborough is a village practically composed of fishermen and their families, and at this small place, with its two miniature bays or " landings," where the cobbles are hauled ashore, quite a large and regular fishing industry exists, the value of which exceeds £6000 per annum. At Bridlington is a small fleet of fishing vessels which put to sea all the year round, and a considerable quantity of fish is sent inland, although the figures do not quite reach those of Flamborough. In the summer months, however, numerous pleasure parties are organised, and are responsible for the capture of quantities of the smaller fish, particulars of which it is not possible to ascertain accurately. At Hornsea fishing in the ordinary sense of the word is carried on to a limited extent, the total annual value of the captures of wet fish being under £200. Against this, however, must be placed the quantity of crabs, obtained by means of "pots," the value of which is something like £600 per annum.

At Hull is one of the principal centres of the fishing industry in the country, and to a large extent the prosperity of the port depends upon its fishing trade. The fishing is carried on by a fleet of about 500 steam trawlers, which work on the Dogger Bank, and other parts of the North Sea, and even extend operations as far as Iceland, the White Sea, to the north and west coast of Scotland, and the coast of Spain. Some of the vessels have been remarkably successful in the White Sea, and have secured catches which realised over £1000 each.

The total value of the fish landed at Hull is approximately £800,000 per annum, and in addition to the thousands of hands employed directly and indirectly in connection with the fishing, there are about sixty smokehouses for curing haddocks, herrings, etc., which give employment to an additional 2000 persons.

The Hull fishing industry is carried on at the St. Andrew's Dock and " Billingsgate," where exceptional facilities are available for landing the traffic, and where there are large areas available for drying and curing. There are also ice factories, cold stores fitted with refrigerators, etc., near the quays where the fish is landed.

After the town of Ravenspurn (which formerly existed at the Humber's mouth) was washed away by the sea in the fourteenth century, the merchants perforce had to find new homes ; some settled at Grimsby and others went to Kingston-upon-Hull, which in those days was a comparatively small port. It had, however, exhibited signs of prosperity since it had secured the royal patronage of King Edward I. In mediæval times, the trade of Hull was of a miscellaneous character, and, though the vessels were only of small size, they sailed to far distant countries. As time went on, trade ran into definite channels, and for considerably over a century was largely centred in the whaling industry, which originated both the oil and fishing trades of to-day. Fleets of sailing vessels went into the Arctic in search of the largest of our mammals, the whale, and considerable sums of money were earned in their capture. In those days whalebone and whale oil were in great demand, and were put to a considerable number of uses before they were superseded. Eventually the introduction of steam enabled voyages to the Arctic to be more expeditiously carried out ; but by that time the whales were becoming so scarce that their search became unprofitable, and in 1863 the last of the whalers sailed from Hull.

Steam also gave an impetus to the shipbuilding trade, and from that time onward Hull has been an important shipbuilding centre. The early steam paddle-boats were quite an institution in the district, and an enormous

JAW-BONES OF WHALE USED AS GATE-POSTS.

passenger and goods trade was carried on between Hull and London, and to the nearer continental ports. Eventually, additions of steamships to the fleets of fishing smacks were made. These were in turn superseded by the modern steam trawlers, of which there are about 500 owned by Hull firms. Along the banks of the River Hull, oil, seed-cake, and flour mills have sprung up, and to-day give employment to a large section of the inhabitants of the city. As shewing the special importance of the oil trade in Hull, out of the 420,000 tons of soya beans imported into the United Kingdom in 1910, over 245,000 tons came through Hull. Similarly Hull had 358,000 of the 690,000 tons of cotton seed, and also large proportions of the linseed, rape seed and castor seed imports. There are likewise extensive timber yards, where, in addition to enormous quantities of sawn timber and telegraph and telephone poles, may be seen hundreds of stacks of pit props waiting for transhipment to the collieries of the West Riding.

Other valuable imports at Hull are grain, fruit and vegetables valued at over £1,000,000 sterling, provisions (butter, eggs, bacon, etc.), £5,000,000 sterling. Much of the wool used in the West Riding is imported at Hull, and at the present time efforts are being made to increase this particular trade.

In the export trade Hull is principally noted for the large quantities of coal sent to all parts of the world, and for milling machinery, much of which is manufactured in the city. It is a convenient place for the export of the textile manufactures of the West Riding. There are several important fleets of steamers at Hull, the most important being that of Messrs. Thos. Wilson, Sons & Co., which connects the port with almost every part of the world. The greatest number of Hull steamers trade with the Baltic, the nearer continental ports, and the Mediterranean ; but some trade direct to places as far distant as Australia.

As a rule the climate of any area principally depends upon its latitude ; though places upon the same latitude may have different climates according to their proximity to the sea, the direction of the prevailing wind, and the relative amounts of sunshine, temperature, and rain. Generally speaking the climate of the British Islands is influenced by the Atlantic Ocean and the Gulf Stream, and sudden changes in the temperature, etc. occur as a result of cyclonic or anticyclonic conditions, which are caused by great disturbances in the atmosphere, which may be compared with the eddies in a stream.

To a certain extent the climate of East Yorkshire is governed by the fact that it is bordered on the east by the North Sea, and protected on the west by the Pennine Chain, which causes a precipitation of a proportion of the moisture from the west before it reaches the riding. These conditions would naturally suggest that the climate of East Yorkshire is exceptionally dry ; and, as a matter of fact, observations made at Patrington, which is towards the south-east corner of the riding, show that the average rainfall there over a period of 52 years is only 23.3 inches. At Hull the average rainfall during 32 years is 27 inches ; Driffield, 26 inches ; and York, 24.2 inches. These places are selected from nine stations in the riding which can show records for over 35 years ; though in all there are 65 stations in the riding which have provided material for Dr. H. R. Mill's excellent rainfall map of East Yorkshire published by the Geological Survey in 1906.

From this map it is clear that a stretch of country averaging 2 miles in width, and stretching from Millington and Bishop Wilton to Thwing and North Burton, receives the greatest amount of rainfall, namely above 32.5 inches per annum. This practically corresponds to the highest part of the Wold district.

A much larger area, though somewhat of the same shape, and extending from Pocklington through North Grimston to Hunmanby, and back again *via* Rudston and Fimber Station to Pocklington, averages between 30 and 32.5 inches. A small oval area near Cherry Burton has the same rainfall. Then a large tract of land as far east as Driffield and Beverley has an average of 27.5 to 30 inches ; whereas a still larger area extending from near York to Brough, and through Hull to a little south of Hornsea, and including Flamborough and Filey, averages between 27 and 27.5 inches. The remainder of the riding, namely the south-east corner as far as Hedon and Aldborough, and the flat country between York and Goole, is below 25 inches. From these particulars, as might be expected, there is a greater rainfall on the Wolds than on the low ground. The average for the whole riding is 27.4 inches. From an economic point of view, it is fortunate that the heaviest rainfall occurs on the chalk area.

The wettest month during the year is October, in which about an eighth of the year's rain falls ; August comes next, during which a tenth of the year's rain falls, this large amount being principally due to heavy thunderstorms which occur in that month.

An interesting piece of confirmatory evidence as to the dryness of the East Riding is shown by the comparative paucity of mosses (which are essentially damp-loving plants), as compared, say, with the North or West Ridings.

With regard to temperature ; observations extending over a number of years at Hull, Driffield, and York, show that the annual average or mean at the first place is 47.6 degrees ; at the second, 46.8 degrees ; whereas at the third, it is 48 degrees.

The average annual sunshine at Hull is 1024 hours ; at Driffield, 1414 hours ; and at York, 1311 hours. The Hull figures, which are somewhat unexpectedly low, may be under-estimated on account of the recording instrument being of an old pattern.

INDEX

315

INDEX

INDEX

317

INDEX

INDEX

INDEX

INDEX

INDEX

INDEX

INDEX

INDEX

325

INDEX

INDEX

INDEX

INDEX

PRINTED AT BROWNS' SAVILE PRESS,
SAVILE STREET AND GEORGE STREET, HULL.

A LIST OF BOOKS

RELATING TO YORKSHIRE

SELECTED FROM THE CATALOGUE OF

A. BROWN & SONS, Ltd.

---

*Further particulars respecting these, and other
similar Works, will be gladly posted
free on application.*

GEOLOGICAL RAMBLES

IN

EAST YORKSHIRE,

By THOMAS SHEPPARD, F.G.S.

247 pages, Demy 8vo , suitably bound in cloth, 3/9 net.

With over 50 Illustrations from Photographs, &c., by GODFREY BINGLEY
and others, and a Geological Map of the District.

CONTENTS.—Introduction—Spurn and Kilnsea—Kilnsea
to Withernsea—Withernsea to Hornsea—Hornsea to Bridlington—Bridlington to Danes' Dyke—The Drifts of
Flamborough Head—South Sea Landing to Speeton—
Speeton and Bempton—The Speeton Clay and Filey Bay—
Filey Brig—Filey Brig to Gristhorpe—Gristhorpe to Scarborough—Scarborough—Scarborough to Robin Hood's Bay
—Robin Hood's Bay—Robin Hood's Bay to Whitby (the
Yorkshire Lias)—Whitby to Redcar—The Humber—Hull
to Hessle—Hessle—Hessle to Brough—The Oolites of Brough
and South Cave—The Yorkshire Wolds—Holderness—Index.

The Hull Daily Mail.—"That East Yorkshire is, for various reasons,
a rich field for the study of geology, scientists have long been aware,
and the exhaustive and instructive work of Mr. Sheppard, illustrated
by photographs, will be welcomed as embodying, in a convenient and
accessible form, much authentic knowledge of the district. The publishers have done their part of the work well."

The Hull E.M. News.—"In the author of this book the reader will
recognise one thoroughly conversant with the field of operations, so
much so that, though so rare and intelligent a companion would be the
greatest of pleasures in an afternoon's outing, yet he has so arranged
his rambles that it is possible even for a stranger to safely walk abroad
without him and yet gain the fullest pleasure and information from the
book alone."

The Naturalist.—"There is not a dull page in Mr. Sheppard's book.
. . . The book ought to find its way into the hands of everyone who
spends a holiday on the Yorkshire coast, while it is still more interesting
to all who dwell in East Yorkshire."

London : A. BROWN & SONS, Ltd., 5, Farringdon Avenue, E.C.
And at Hull and York.

THE EVOLUTION OF KINGSTON-UPON-HULL

AS SHEWN BY ITS PLANS.

By THOMAS SHEPPARD, F.G.S., F.S.A.(SCOT.),

Curator Municipal Museums, Hull.

*204 pages, Demy 8vo, fully illustrated and with Copious Index.
Bevelled Cloth Boards, 3/6 net.*

This work has been compiled at the request of the Museums and Records Committee of the Hull Corporation. I have been kindly permitted to examine various plans at the British Museum, and in numerous public and private collections, the particulars of which can be gathered by a perusal of the list contained in the book. I trust that as a result the book contains not only a complete list of all the plans of any importance relating to the City, but also a record of the valuable information they convey. Incidentally, the rise and progress of Kingston-upon-Hull is recorded as the plans, one by one, are described.

As it has been possible to bring forward particulars of some plans, etc., not previously known to exist, I am pleased to think that in this small volume will be found some new light upon the history of the third port.

THE MAKING OF EAST YORKSHIRE.

A CHAPTER IN LOCAL GEOGRAPHY.

By THOMAS SHEPPARD, F.G.S., F.S.A.(SCOT.).

29 pages, Demy 8vo, illustrated with 4 full-page plates on Art Paper, and strongly bound in Stout Covers, 1/-net.

Nature Notes.—" It is certainly a good notion that Teachers should be given a simple elementary exposition of the geological structure and history of their district, illustrated with views, which they can cut out and pin up on the blackboard. Mr. Sheppard is fully competent to provide such a *resume* of the geology of East Yorkshire."

London : A. BROWN & SONS, Ltd., 5 Farringdon Avenue, E.C.
And at Hull and York.

THE LOST TOWNS OF THE HUMBER

WITH AN INTRODUCTORY CHAPTER ON THE ROMAN GEOGRAPHY OF SOUTH EAST YORKSHIRE.

By J. R. BOYLE, F.S.A.

114 pages Royal 4to, Map and Folding Pedigree.
Cloth Boards, 7/- net.

CONTENTS:—Roman Times—Saxon Times—Ravenser—Chapel of Ravenserodd—The Burgesses and Land.Owners of Ravenser—Property in Ravenser—Ravenser Spurn—The Sites of Ravenser and Ravenserodd —Tharlesthorp — Frismersk — Sunthorp—Penisthorp—Orwythfleet — Appendices and Index.

Yorkshire Notes and Queries.—"The work fills a place hitherto vacant in Yorkshire Topography, and will be sought after for ages."
Hull News.—"The volume will take a place among standard works. . . . Materials have been drawn from unpublished MSS. in the Public Record Office and British Museum."

NOTES RELATIVE TO
THE MANOR OF MYTON.

By J. TRAVIS-COOK, F.R.H.S.

One volume, cloth bound, uncut, Demy 8vo, 4/11 net.
Large paper, Demy 4to, 9/9 net. Also a limited edition of the latter on hand-made paper, 15/6 net.

Illustrated with a fac-simile from the Doomsday, and three Maps.

CONTENTS:—Of Manors Generally — Etymology of the Name "Myton"—History of Manor of Myton—Manor of Tupcoates, with Myton—The Change in the Course of the River Hull.

Yorks. County Magazine.—"From first to last these pages teem with solid information, and the book must take its place as a standard work of reference on Myton and the district around Hull."

London : A. BROWN & SONS, Ltd., 5 Farringdon Avenue, E.C.
And at Hull and York.

FORTY YEARS' RESEARCHES

IN

BRITISH & SAXON BURIAL MOUNDS

OF

EAST YORKSHIRE

INCLUDING ROMANO-BRITISH DISCOVERIES, AND A
DESCRIPTION OF THE ANCIENT ENTRENCHMENTS
ON A SECTION OF THE YORKSHIRE WOLDS.

BY J. R. MORTIMER

(Founder of the Mortimer Museum at Driffield),

WITH OVER 1000 ILLUSTRATIONS FROM DRAWINGS BY

AGNES MORTIMER.

*800 pages, 12 × 8, bound in a Seal Back, Cloth Sides, Gilt
Top, 50s. net.*

Extract from " The Times " Review.

" The volume is chiefly occupied with an account carefully written
down in each case at the time and on the spot, of the opening and
examination of about 350 British burial mounds, mostly of the ' round
barrow ' type, and of some later Anglo-Saxon cemeteries, accompanied
by diagrams showing in plan and section the situations of the bodies
therein interred, and the positions—often a matter of importance—
and state of preservation of the bones. The latter information is
afforded by spirited little delineations of the skeletons, drawn from
sketches made at the time by Mr. Mortimer. The funeral furniture
of the graves—consisting of pottery almost always ornamental, with
linear patterns, flint and stone implements and weapons, and more
rarely of similar utensils of bronze and iron ; of objects of personal
adornment and others of a miscellaneous kind—is represented partly
in the text, but mainly in 125 half-tone plates (in which are shown
more than 1000 different objects) from drawings by Miss Agnes Mortimer. The accuracy of these drawings is guaranteed by a competent
authority, who has compared each one of them with the originals.
Several problems of great archæological interest are suggested by Mr.
Mortimer's book. One of these is the relation to the burial mounds
of the defensive earthworks, which are so numerous in this same district
of the Yorkshire Wolds ; and another is the significance of certain
cruciform excavations and works in which the author sees a Christian
reference. For the working out of these and other problems, Mr.
Mortimer supplies a body of carefully ascertained facts, for which all
subsequent investigators and theorists will owe him a debt of gratitude."

London : A. Brown & Sons, Ltd., 5 Farringdon Avenue, E.C.
And at Hull and York.

THE BIRDS OF YORKSHIRE

Being a Historical Account of the Avi-Fauna of the County,

By T. H. NELSON, M.B.O.U.,

With the co-operation of W. EAGLE CLARKE, F.R.S.E., F.L.S., and F. BOYES.

There has scarcely been a Yorkshire Naturalist living within the past 35 years who has not contributed manuscript notes or lists to the store available for reference.

Small Paper Edition.—Demy 8vo, containing 901 pages of letterpress and upwards of 200 illustrations from photos by R. Fortune, F.Z.S., and other well-known naturalist photographers, beautifully printed in double tone ink on best Art Paper, also 3 three-colour plates and specially designed title pages in colours, strongly bound in a fast coloured cloth binding. 25/- net.

Large Paper Edition (only 250 copies printed).—Demy 4to, specially prepared with wide margins for additional records, notes, &c. 42/- net.

The famous Naturalist Lecturer, Mr. R. KEARTON, F.Z.S., in his review of this work for the " Daily Chronicle," wrote :—" In his preface the author says that this work is based upon an unrivalled and exceptionally complete mass of material, that it is comprehensive in scope, and that the account of each species dealt with in its pages includes particulars of faunistic position, distribution, migration, nidification, folk-lore, varieties, and vernacular names. This bold claim is thoroughly justified. Having been born and brought up amongst the birds in one of the wildest parts of the 'County of Broad Acres,' I felt myself more or less qualified to test the accuracy of the author's statement when his two handsome volumes came into my possession, and I am bound to confess that he does not in the least overstate the claims of the work. I have again and again put its accuracy and fulness to the severest of tests, and am bound to confess that in each instance it has come out triumphant. Mr. Nelson and his literary and pictorial helpers have placed all British ornithologists under a deep debt of gratitude by the production of one of the best and completest county histories of birds ever published."

The Yorkshire Post.—" Never was a county monograph undertaken with more efficient leadership and co-operation, and never, it may be added, has one been compiled that will have more lasting value. It is pure delight from beginning to end."

The Scotsman.—" It appears as the bird history of the county."

The Westminster Gazette.—" The illustrations are numerous and good, and the book is altogether excellently turned out."

London : A. BROWN & SONS, Ltd., 5, Farringdon Avenue, E.C.
And at Hull and York.

YORKSHIRE FOLK-TALK

WITH

CHARACTERISTICS OF THOSE WHO SPEAK IT IN THE NORTH AND EAST RIDINGS

BY THE

Rev. M. C. F. MORRIS, B.C.L., M.A.,

Late Rector of Nunburnholme, Yorkshire.

---

458 pages, Crown 8vo, Strongly Bound in Cloth Boards, with Gilt Top. **4/6** *net.*

SECOND EDITION

With an Addendum to the Glossary.

EXTRACT FROM PREFACE.

Although the first edition of *Yorkshire Folk-Talk* met with so favourable a reception from the public, and the issue was so soon exhausted, it is only now, after a lapse of nineteen years, that I have found it practicable to put forth a new and cheaper one.

During the past twenty years we find that the broad Yorkshire talk of former days has, in most districts, suffered loss from contact with that of the world outside. The old folks pass away from us, and those of another, though not better tongue, take their place. But this fact should only make us cherish the more that which still is left to us. And assuredly, whether the language of the people of East Yorkshire be living or dead, it will well repay careful study. The more closely we examine it, the more interesting it becomes.

London: A. Brown & Sons, Ltd., 5 Farringdon Avenue, E.C.
And at Hull and York.

YORK IN ENGLISH HISTORY.

A spirited and zestful book, profusely illustrated, and of fascinating interest to all lovers of history.

By J. L. BROCKBANK, M.A., AND W. M. HOLMES.

304 pages, Crown 8vo, with 58 Illustrations, artistically bound, with City Arms in gold and colours on side, gilt top 3/- net.
Cheaper Edition, on ordinary paper, cloth bound 1/8 net.

HIGH PRAISE FROM THE PRESS.

" From beginning to end the authors hold the reader's attention by their clear and vigorous style. The chapters on the Roman period, the various Scottish raids, the early mystery plays, and the '45 Rebellion are particularly good, while that on the Pageant of Mediæval York makes one wish that it were longer. Old coaching days are lovingly treated, and elections in the 'good old times' are well described in a manner reminiscent of the famous Eatanswill election."

" We have nothing but praise for this charming book. It has been well said that 'to master thoroughly the story of the city of York is to know practically the whole of English history,' and the authors of this new history have demonstrated the truth of this opinion. From that almost prehistoric time when the Celts settled in *Eburach*—the field at the meeting of the waters—through the Roman occupation and fortification of *Eburacum*, and on through the Anglian development of *Eoferwik*, and the Danish colonisation of *Jorvik*—i.e., Yorwick—we are led on to the *York* of Norman times, and so through mediæval ups and downs to the city as we know it to-day. Across its stage have passed Julius, Agricola, Hadrian, Severus, and Constantine; Edwin, Siward, Tostig, Harold; William the Conqueror and Edward, *Malleus Scotorum*, Queen Phillipa and the fair Margaret, James I., and all the Stuart kings; Fairfax and Cromwell, and the gay, dashing Cavalier, Prince Rupert. And parallel with these there have been the leaders of religious thought, and the grand, old Minster, looking calmly down on scenes of war and Revolution. This is the story that is told so admirably by Messrs. Brockbank and Holmes—a story which no resident in, or visitor to York should leave unread. No pains have been spared by the publishers to give the letterpress a perfect setting; binding, paper, illustrations, and general finish are alike admirable."

London: A. BROWN & SONS, Ltd., 5 Farringdon Avenue, E.C.
And at Hull and York.

THE STORY OF THE
EAST RIDING OF YORKSHIRE

By HORACE B. BROWNE, M.A.

*368 pages Crown 8vo, printed on Art Paper and bound in Art
Cloth Boards with 170 illustrations.
3/- net, or free by post for 3/4 net.*

The book is one that can be read with profit by both
young and old. It is a work that will impress upon
Yorkshire people the wealth of Archæological, Architectural,
Civic and Commercial matter which lies in their midst. In
fact it should make every journey in the East Riding full
of interest and pleasure.

The abundance of illustrations in the book—photo-
graphic reproductions and specially prepared maps and
diagrams, many of which are here published for the first
time—should alone serve to make it a popular one.

The Eastern Morning News.—"The story of this particular part
of the county has been told before, it is true, but we question if
anything so lucid and so readable has ever been brought together
regarding the Riding in one volume. Mr. Browne clothes his subject
with an interest that is nothing short of fascinating. His book is all
plain and pleasant sailing; there is a clever avoidance of the rocks
of antiquarianism. What antiquarian lore is served up to us is so
admirably treated, so free from dry-as-dust forms of presentation, that
the title of 'story' is fully justified. The reader under Mr. Browne's
guidance glides through the sea of ancient history knowing the antiquarian rocks are there, but never comes into too close contact with
them. We emphasise this point because as a rule, the general reader
fights shy of a book on such a subject as this, fearing he may be bored."

The Yorkshire Post.—"'The Story of the East Riding,' by Mr.
Horace B. Browne, M.A., brings into entertaining narrative a great
mass of historical material, ancient and modern, and it is all very well
illustrated. One of its best chapters is a description of life in a mediæval
town—its commerce, popular amusements, and so on; a well-written
history of Hull follows; we then have a picture of the old coaching
days, and the introduction of railways, gradual but revolutionary,
some notes on famous men of the Riding (also ancient and modern),
and a study of the folk-speech of East Yorkshire, compared with Old
English and standard English works. It is altogether a carefully
written work."

London: A. BROWN & SONS, Ltd., 5 Farringdon Avenue, E.C.
And at Hull and York.

THE

REGULATIONS & ESTABLISHMENTS

OF THE HOUSEHOLD OF

HENRY ALGERNON PERCY,

THE FIFTH EARL OF NORTHUMBERLAND,

AT HIS CASTLES OF

WRESSLE AND LECKONFIELD

IN YORKSHIRE.

BEGUN ANNO DOMINI MDXII.

---

*A New Edition.   Edited with Additional Notes.
488 pages, Demy 8vo, Uncut, Gilt Top, strongly bound in
Art Canvas and Stout Boards to match.   Price 8/6 net.*

THIS "Northumberland Household Book" is a faithful
reprint of the first edition, edited and annotated from the
original MS. by Bishop Thomas Percy in 1770.   For this
edition, collation with the original MS. was unfortunately
found to be impracticable.   There is, however, little reason
to doubt the accuracy of Bishop Percy's text.   Indeed,
that text bears internal evidence of having been very
carefully prepared.   Percy's Preface is reprinted, without
alteration, in the present volume.   Considerable additions,
however, have been made to Percy's notes.   Obscure
allusions, on which he made no comment, have been
elucidated, and his annotations on many points have been
very greatly expanded.   But nothing has been omitted
from his notes, and all the additions, for which the
present editor is solely responsible, have been enclosed in
square brackets.

London: A. BROWN & SONS, Ltd., 5, Farringdon Avenue, E.C.
And at Hull and York.

THE FLORA OF THE EAST RIDING OF YORKSHIRE,

Including a Physiographical Sketch.

By JAMES FRASER ROBINSON.

With a List of the Mosses, By J. J. MARSHALL.

And a Specially Prepared
Coloured Geological Map, showing the Botanical Divisions of the District.

253 pages, Demy 8vo., Bound in Cloth Boards, 3/9 net.

A special Interleaved Edition has also been prepared for notes, 7/6 net.

Although almost every county in England has its published "Flora," and the plants of the North and West Ridings have been described by Mr. J. G. Baker and Dr. F. A. Lees respectively, hitherto no "Flora" of the East Riding has been issued. The present work consequently supplies a want long felt, not only by field naturalists and scientific men in general, but by all who are interested in the country's flora. The author has for 17 years been carefully studying the plants of the East Riding, and has also compiled from all possible sources anything pertaining to the plant inhabitants of the vice-county. He has also been assisted during that period by the Members of the Hull Scientific and Field Naturalists' Club, the weekly field excursions of which, into all parts of the Riding, he has rarely missed.

The Journal of Botany.—"British botanists will find much information in this volume, and will do well to place it on their shelves."

Nature.—"The Author and the Hull Scientific and Field Naturalists' Club deserve the thanks of botanists for a compilation which represents much hard work, and which will serve to stimulate interest in that division of the county, inasmuch as it indicates a somewhat unexpected wealth and variety of plant forms."

Knowledge.—"Among the many local floras published of late years, the present book will take a high place."

London: A. BROWN & SONS, Ltd., 5, Farringdon Avenue, E.C.
And at Hull and York.

THE EARLY HISTORY

OF THE

TOWN AND PORT OF HEDON

IN THE COUNTY OF YORK.

BY J. R. BOYLE, F.S.A.

495 pages, uncut edges, Demy 8vo, half-bound and gilt top, 21s. net.

A few hand-made paper copies, Demy 4to, bound in buckram and gilt top, 42s. net.

" Trueth is that when Hulle began to flourish, Heddon decaied."—*Leland.*

CONTENTS:—The Origin of Hedon—The Borough of Hedon—The Port of Hedon—The Churches of Hedon—The Institutions of Hedon—The Topography of Hedon—Tenure of Hedon—An Appendix of 34 Notes, occupying 207 pages—Glossary—Index.

The little town of Hedon has a history of almost unique interest. Springing into existence very soon after the Norman Conquest, it rapidly advanced to a position of wealth and prosperity, which ranked it, in the twelfth century, amongst the great towns of England. But before the nineteenth century its decline had commenced, and from that period, despite many spasmodic attempts to revive its ancient *prestige*, its importance has slowly and gradually ebbed away, until now, to the casual observer, there is little except its fine church to distinguish it from any other quiet English village of its size.

In the present work the origin of Hedon, its progress, and the early period of its decline, are carefully traced. Its institutions; civil and ecclesiastical, are fully described. The illustrations include a Plan of the Town, a fac-simile of the Charter of King John, and engravings of the Ancient Maces and Seals.

The appendix contains full transcripts of the documents on which the writer has based his history of the town. These include the whole of the Regal Charters granted to Hedon from that of Henry II. to that of Henry V.; early *compóti* of the Bailiffs and Chamberlain, of the Wardens of the Churches of St. Augustine, St. Nicholas, and St. James, and of the Warden of the Chantry of St. Mary; Rentals of the Manor; Extracts from Court Rolls; many important Land Charters; and the Constitutions of the Borough, *temp* Philip and Mary.

London: A. BROWN & SONS, Ltd., 5, Farringdon Avenue, E.C
And at Hull and York.

YORKSHIRE MOORS AND DALES.

A Description of the Moors of North-East Yorkshire, etc.

By ALFRED P. WILSON.

In one Volume. Size 8¾ by 6¼ inches, tastefully bound in Cloth Boards, lettered in gold, with gilt top ; contains 236 pages and 12 full-page plates on Art paper. 10/6 *net.*

BOOK I.—A Short Guide to the Moors.

The Moors of the North West—Northern and Central Moors—Southern Uplands—The Hambletons —Eastern Moors.

BOOK II.—Yorkshire Moors and Dales.

Unexplored Yorkshire—Old-Fashioned Roofs— " What Mean These Stones ? "—Lost on the Moors— Moorland Roads—About the Dalesfolk—The Old Dalesman—Old Manners and Customs—Lightning and Tempest—Farming in the Dales—The Waggoner— Wild Nature—Grouse Shooting—Dialect, Place Names and Glossary—The Ill-natured Person—Musings in the Dales—Gleanings.

BOOK III.—Tales and Sketches Descriptive of the Dalesfolk.

Liverpool Daily Post :—" A charmingly written and beautifully illustrated volume, which will serve not only as a guide to the moors, but also as an incentive to antiquarians and travellers, and as an enjoyment to all who study mankind, and his habitat. The book, which is well printed, with wide margins, should be in every Yorkshire-man's library."

Sheffield Daily Telegraph :—" Our reading this week-end included a delightful book on the moors and dales of North Yorkshire. It deals with a district we have several times visited. . . . Taking it all round, this is a most interesting and enjoyable book, which we can heartily recommend."

Yorkshire Post :—" This book, and the ordnance sheets covering the Cleveland Hills, should be in the hands of everybody who contemplates making personal acquaintance with the district. The illustrations are good, and the information sound."

London : A. Brown & Sons, Ltd., 5 Farringdon Avenue, E.C. And at Hull and York.

ANDREW MARVEL

AND HIS FRIENDS.

(A Story of the Siege of Hull),

By MARIE HALL.

Ninth Edition, containing 485 pages, Crown 8vo., in a characteristic binding specially designed by J. Walter West, 3s. 6d.

FROM no book hitherto written can the reader gather a more vivid or accurate conception of events which characterised the two Sieges of Hull than he will derive from this volume. Not less striking and faithful are the Author's pictures of the English Court as it existed both during the Protectorate and the reign of Charles the Second. It is hitherto the only piece of historical fiction, the chief scenes of which lie in Kingston-upon-Hull. Hull, indeed, with its stirring history and its wealth of ancient tradition, its unbroken line of princely merchants, stretching from the time of De la Poles, to that of Lister and Raikes and Thornton and Wilberforce, afforded a new quarry upon which Mrs. Hall seized, and she has told her story well.

The London Daily Telegraph.—"At a time when so much trash is poured out upon the public, a volume of pure and sweet sentiment like this should be heartily welcomed."

London : A. BROWN & SONS, Ltd., 5, Farringdon Avenue, E.C. And at Hull and York.

ACROSS THE BROAD ACRES.

BEING SKETCHES OF YORKSHIRE LIFE & CHARACTER.

BY

Rev. A. N. COOPER, M.A.,

VICAR OF FILEY, YORKS.

Author of " With Knapsack and Note-Book." "Quaint Talks about Long Walks," etc.

With 8 Full Page Photo-Illustrations on Art Paper.

728 Pages, Crown 8vo, Tastefully Bound in Art Vellum Boards. **3/6** *net.*

The Broad Acres of Yorkshire have been productive of useful men, strenuous men, and men of shrewd, sound common sense, but it is curious that the county has never produced a poet or songster of the first rank,   There seems to be something in the air which represses the imagination, and the people are practical to a degree.

This will account for the practical, but I trust none the less interesting nature of much that is contained in this book. My life has been mainly passed between the Waves and the Wolds, and it is best for all, except those who possess the gift of imagination, to write about what they know. The following chapters deal chiefly with the men and things of East Yorkshire.

The Scotsman—" Highly entertaining and pleasantly blended with learning and culture. Altogether a delightful book."

The Daily Chronicle—" A capital antidote to a fit of the blues.' '

The Yorkshire Observer—" Readers will find it full of entertainment."

The Yorkshire Post—" The chapters are always chatty, jocular, and serious by turns, as a walking companion ought to be."

The Newcastle Chronicle—" The Walking Parson has a gift for painting scenery in words, and treats us to some thoroughly amusing stories."

The Daily News—" The writings of the Walking Parson have a charm peculiarly their own."

The Spectator—" Altogether a very pleasant book."

London : A. BROWN & SONS, Ltd., 5 Farringdon Avenue, E.C.
And at Hull and York.

ROUND THE HOME

OF A

YORKSHIRE PARSON:

OR

STORIES OF YORKSHIRE LIFE.

By the REV. A. N. COOPER, M.A.,

VICAR OF FILEY, YORKS.

Author of " Quaint Talks about Long Walks."

Cheap Edition, revised and containing some entirely new Chapters, 320 pages, Crown 8vo, with 8 full-page Plates.

1/- net.

THIS most interesting volume will be found particularly suitable for holiday reading, and should commend itself to all lovers of East Yorkshire. A vein of humour runs through all the stories that at once makes them fascinating and amusing.

The Morning Post.—"Such books are not to be read every day of the week, nor are many parsons so well worth hearing. It will be our own fault and our own loss if we do not seize this opportunity of making the acquaintanceship of the Parishioners of Filey through the pages of their broad-minded and generous-hearted Vicar."

The Yorkshire Post.—"A book of the most entertaining kind, and one, too, that is worth taking home and placing on one's bookshelf."

The British Weekly.—"I can cordially recommend 'Round the Home of a Yorkshire Parson.' It is really a fine, manly book, frank, cheerful, and by no means without literary power. Yorkshire people and the multitudes who love Yorkshire, with good reason, will like it, and they will have no hesitation in pronouncing the author emphatically a good fellow."

London: A. BROWN & SONS, Ltd., 5, Farringdon Avenue, E.C.
And at Hull and York.

A REEL OF No. 8

AND

SUDDABY FEWSTER

(Two Holderness Tales),

By FLIT and KO.

Illustrated by J. WALTER WEST and others.

New Edition, Bound in Cloth with Gilt Top and Rough Edges, 3s. 6d. net.

LIST OF ILLUSTRATIONS.

Frontispiece, Ploughing Scene, by Walter West.—"What a beautiful head of hair you've got!"—"Ah seed you arming old gel across closes last night."—"An' then there was that uncomfortableness aboot jacket-waast."—"If Ah gets a wife, why, Ah *diz ;* an' if Ah dizen't get one, why, Ah *dizn't.*—Suddaby Fewster.—"'Then bloonder in, lad,' I says." —"I wants Little Un to get them two pot dogs on chimney-piece."— "It's a very useful thing is a black corran."—"Here's a parcel for yer as came this morning."

The Bradford Chronicle.—"Whoever 'Flit and Ko' may be, their delightful book is something to be truly thankful for. It is becoming trite to call every dialect writer the 'Barrie' of his particular district, but assuredly 'Flit and Ko' run no danger of being outshone even when compared with the writer of 'Thrums' and the creator of 'Jess.'"

Pall Mall Gazette.—"It is a refreshing bit of simple life to come upon in the wilderness."

Sheffield Daily Telegraph.—"To lovers of Yorkshire and lovers of nature these two tales may be confidently commended. They are studies from life, carefully rendered, and with hardly a weak touch throughout."

Manchester Guardian.—"The dialect is wisely simplified, and the authors know it as well as they know the English of Literature. It leaves upon the mind a vivid and picturesque impression."

Eastern Morning News.—"The sturdy independence, abrupt manner, and the keen sense of humour which form part and parcel of the generous and kind-hearted disposition of the Holdernessians, have been most faithfully portrayed, The whole story is instinct with life, fascinating in originality, freshness, and sympathetic treatment."

London: A. Brown & Sons, Ltd., 5, Farringdon Avenue, E.C.
And at Hull and York.

How to be Happy though Hunted, from the Foxes' Point of View:
A charming story written in the East Yorkshire Dialect. By FLIT.
92 pages, sewn in an Artistic Cover. **1/- net.**

A Holderness Harvest: An East Yorkshire Dialect Story. By FLIT.
150 pages and 4 full-page illustrations from Author's own sketches,
sewn in an Artistic Cover. **1/3 net.**

**Brazzock, or Sketches of some Humourous characters of a Holderness
Parish.** By the Rev. WILLIAM SMITH, Rector. 235 pages, Crown
8vo, Illustrated, bound in Art Vellum. **3/6 net.**

The New Copyright Official Handbook to the City of Hull. By Sir
ALBERT KAYE ROLLIT, LL.D., D.C.L., D.Litt. Beautifully printed
on Art Paper, and bound in Art Vellum Boards. **2/6 net.**

**Evidences relating to the Eastern Part of the City of Kingston-upon-
Hull.** By THOS. BLASHILL, F.R.I.B.A. **2/6 net.**

Essays upon the History of Meaux Abbey, and some Principles of
Mediæval Land Tenure. By the Rev. A. EARLE, M.A. **3/6 net.**

The Birds of Bempton Cliffs. By E. W. WADE, M.B.O.U. With 18
Illustrations from Photographs taken by the Author. A concise
and interesting History. Price **1/- net.**

**Lines Written on a Visit to Wilberforce House, Hull, and other Verses of
Local Interest.** By FRANK NOBLE WOOD. Dedicated to the
Mayoress of Hull. 64 pages, printed in best style on Antique
Paper, with appropriate cover, price **1/- net.**

A History of South Cave and other Parishes in the County of York. By
J. G. HALL. 8vo, **5/- net**; 4to, **10/- net.**

Church Bells of Holderness. By G. R. PARK. Crown 8vo, cloth. **1/6 net.**

**The History of God's House of Hull, commonly called the Charterhouse,
from its Foundation.** By J. TRAVIS-COOK. **7/6 net.**

Fac-simile Reprint of the First Hull Directory, Published in 1791.—
Containing Names, Professions, and Residences. **3/6 net.**

Gents' History of Hull.—Reprinted in fac-simile of the original of 1735,
with folding plans and illustrations. **3/9 net.** Large paper, **5/- net.**

Kingstoniana.—Historical Gleanings and Personal Recollections. By
Alderman JOHN SYMONDS, **7/6 net.**

Speech of Holderness and East Yorkshire. By W. H. THOMPSON, **1/6 net.**

The Poetical Works of Andrew Marvel. With Memoir. **1/6 net.**

Hull Letters.—Printed from the Hull Borough Archives. Period 1625-
1646 (Charles I. until his imprisonment). **3/6 net.**

**Notes on the Charter, Granted by King Edward the First to Kingston-
upon-Hull, 1st April, 1299,** with a translation and illustrated
documents, by J. R. BOYLE, F.S.A. **1/- net.**

St. Patrick's Church, Patrington: its History and Architecture. By
the Rev. Canon MADDOCK. **6d. net.**

"The Naturalist." A monthly illustrated journal of Natural History.
Edited by T. SHEPPARD, F.G.S., and T. W. WOODHEAD, Ph.D.
The oldest Scientific Periodical in the British Isles, dating back to
1833. Subscription price **6/6** per annum post free.

4

London: A. BROWN & SONS, Ltd., 5 Farringdon Avenue, E.C.
And at Hull and York.